T0213660

Simula SpringerBriefs on Computing

Volume 9

Springer and Simula have launched a new book series, *Simula SpringerBriefs on Computing*, which aims to provide introductions to select research in computing. The series presents both a state-of-the-art disciplinary overview and raises essential critical questions in the field. Published by SpringerOpen, all *Simula SpringerBriefs on Computing* are open access, allowing for faster sharing and wider dissemination of knowledge.

Simula Research Laboratory is a leading Norwegian research organization which specializes in computing. The book series will provide introductory volumes on the main topics within Simula's expertise, including communications technology, software engineering and scientific computing.

By publishing the *Simula SpringerBriefs on Computing,* Simula Research Laboratory acts on its mandate of emphasizing research education. Books in this series are published only by invitation from a member of the editorial board.

More information about this series at http://www.springer.com/series/13548

Yan Zhang

Mobile Edge Computing

Yan Zhang
Department of Informatics
University of Oslo
Oslo, Norway

ISSN 2512-1677 ISSN 2512-1685 (electronic)
Simula SpringerBriefs on Computing
ISBN 978-3-030-83943-7 ISBN 978-3-030-83944-4 (eBook)
https://doi.org/10.1007/978-3-030-83944-4

This Springer imprint is published by the registered company Springer Nature Switzerland AG
The registered company address is: Gewerbestrasse 11, 6330 Cham, Switzerland

Preface

This book offers comprehensive, self-contained knowledge on Mobile Edge Computing (MEC), a very promising technology for achieving intelligence in next-generation wireless communications and computing networks. With the rapid development of beyond 5G/6G and the Internet of Things, increasing number of smart devices are being deployed at the edge of networks. Due to the enormous amount of data and long transmission distances, centralized computing mechanisms incur high latency and network congestion. By placing the computing and storage resources closer to the users, MEC can significantly increase performance in terms of low latency, reduced communications overhead, and high-quality user experience. Further, processing data on edge devices will enhance system security and data privacy.

This book allows for easy cross-referencing owing to the broad coverage on both the principle and applications of MEC. It covers the basics, essential topics, and future directions of MEC. It also details the design and implementation of communications, computing, and caching in MEC. The book starts with the basic concepts, key techniques, and network architectures of MEC. Then, we present the wide applications of MEC, including edge caching, 6G networks, the Internet of Vehicles, and unmanned aerial vehicles. In the last part, we present new opportunities when MEC meets blockchain, artificial intelligence, and distributed machine learning (e.g., federated learning). We also identify the emerging applications of MEC in a pandemic, the Industrial Internet of Things, and disaster management.

The objectives of this book are to provide the basic concepts of MEC, to explore the promising application scenarios of MEC integrated with emerging technologies, and to give insights into the possible future directions of MEC. For better understanding, this book also presents a few use cases of MEC models and applications in different scenarios. The primary audience includes senior undergraduate and postgraduate students, educators, scientists, researchers, engineers, innovators, and research strategists. This book is mainly designed for academics and researchers from both academia and industry who are working in the field of wireless networks and edge intelligence. Students majoring in computer science, electronics, and communications will also benefit from this book. The content of this book will also be useful for senior undergraduate students, graduate students, and faculty working in MEC.

Oslo, Norway   Yan Zhang

Acknowledgements

This book was mainly written during the COVID-19 period, and the writing took much longer than expected. My gratitude goes out to all of my excellent students and research collaborators. I appreciate all their contributions of time, discussions, and ideas that made this book possible. Our joint publications in IEEE journals and conferences provided solid, high-quality material for the book.

Special thanks go to Prof. Aslak Tveito at Simula Research Laboratory, Prof. Olav Lysne at the Simula Metropolitan Center for Digital Engineering, and Rachel Thomas at Simula Research Laboratory for their patience and support since the beginning until the final stage. I am very thankful for Simula Research Laboratory, where I worked from 2006 to 2016. This was the most important period of my career development. With the strong support of Simula Research Laboratory and the trust of Prof. Olav Lysne, I was very lucky to receive awards and honors in recent years, including IEEE Fellow and Highly Cited Researcher. The most important scientific contributions that elevated me to IEEE Fellow were carried out at Simula Research Laboratory.

I am very grateful to the staff at Springer for their great efforts during the typesetting period. Last but not least, I want to express my deep thanks to my families and friends for their constant encouragement, patience, and understanding throughout this project during the COVID-19 period.

February 2021   Yan Zhang

Contents

Acronyms

5G	Fifth-generation mobile networks
6G	Sixth-generation mobile networks
AI	Artificial intelligence
API	Application programming interface
BS	Base station
D2D	Device to device
DDPG	Deep deterministic policy gradient
DPoS	Delegated proof of stake
DRL	Deep reinforcement learning
IIoT	Industrial Internet of Things
IoT	Internet of Things
IoV	Internet of Vehicles
LTE	Long-Term Evolution
MBS	Macro base station
MCC	Mobile cloud computing
MEC	Mobile edge computing
ML	Machine learning
NOMA	Non-orthogonal multiple access
OMA	Orthogonal multiple access
P2P	Peer to peer
PBFT	Practical Byzantine fault tolerance
PoS	Proof of stake
PoW	Proof of work
QoE	Quality of experience
QoS	Quality of service
RAN	Radio access network
RSU	Roadside unit
SINR	Signal-to-noise-plus-interference ratio
TDMA	Time division multiple access
UAV	Unmanned aerial vehicle
V2I	Vehicle-to-infrastructure
V2R	Vehicle-to-RSU

V2V	Vehicle-to-vehicle
VEC	Vehicle edge computing
VR	Virtual reality

Chapter 1
Introduction

Abstract This chapter first introduces the fundamental concepts of mobile cloud computing. The differences between mobile edge computing and mobile cloud computing are then discussed in detail. The European Telecommunications Standards Institute's concept of mobile edge computing is introduced with respect to mobile edge computing's definition, architecture, advantages, and potential applications.

1.1 Mobile Cloud Computing (MCC)

MCC integrates cloud computing with mobile devices to enhance mobile device capabilities such as computing and storage. The user experience is improved by the execution of computation- and storage-sensitive applications through cloud computing and the delivery of related services. The architecture of MCC and mobile edge computing (MEC) is illustrated in Fig. 1.1. Mobile devices connect to the web services through nearby base stations. Web services act as the application programming interface (API) between mobile devices and the cloud and deliver cloud applications to the mobile devices. In current architecture, mobile devices can access cloud services through base stations in mobile network or Wi-Fi access points. MCC enables resource-limited mobile devices to run applications that are latency insensitive but computation intensive.

However, the inherent limitation of MCC is the long transmission distance between mobile devices and the cloud, which incurs long execution latencies and cannot satisfy the time constraints of latency-critical applications. There exist significant differences between MEC systems and MCC systems. MEC integrates cloud computing into mobile networks to provide computing and storage capabilities to end users at the edge. The main differences between MCC and MEC are summarized in Table 1.1.

- *Physical server*: In MCC systems, the physical servers are located in large-scale data centers. The data centers are large specific buildings. The buildings are equipped with adequate power supply and cooling equipment. The MCC servers are equipped with high computing and storage capabilities. In MEC systems, however, the servers are colocated with small-scale data centers, such as wireless

© The Author(s) 2022
Y. Zhang, *Mobile Edge Computing*, Simula SpringerBriefs on Computing 9,
https://doi.org/10.1007/978-3-030-83944-4_1

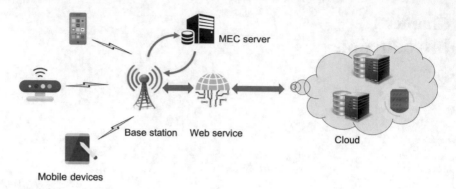

Fig. 1.1 The architecture of MCC and MEC

Table 1.1 Comparison of MCC and MEC

	MCC	MEC
Physical server	High computing and storage capabilities, located in large-scale data centers	Limited capabilities, colocated with base stations and gateways
Transmission distance	Usually far from users, from kilometers to thousands of kilometers	Quite close to users, from tens to hundreds of meters
System architecture	Sophisticated configuration, highly centralized	Simple configuration, densely distributed
Application characteristics	Delay tolerant, computation intensive, e.g., Facebook, Twitter	Latency sensitive, computation intensive, e.g., autonomous driving, online gaming

routers, base stations, and gateways. The MEC servers are equipped with a limited amount of computing and storage resources.

- *Transmission distances*: The distances between MCC servers and users can vary greatly, from kilometers to thousands of kilometers, even encompassing different countries, whereas the distances between MEC servers and end users are usually short, from tens to hundreds of meters.
- *System architectures*: The MCC systems are usually deployed by giant information technology (IT) companies such as Google and Amazon. The architectures of MCC systems are usually very sophisticated and highly centralized. The servers are controlled and maintained by specialized technical individuals. In MEC systems, the servers are usually deployed by telecommunications operators, enterprises, and communities. Theses servers are densely distributed in the network, with a simple configuration. MEC systems are hierarchically controlled in a centralized or distributed manner.
- *Application characteristics*: The applications in MCC systems can usually tolerate a certain degree of latency but require large amounts of computational resources. The computation data can thus be transmitted from end users to the MCC servers for computation. Typical examples of MCC applications are online social networking,

such as Facebook and Twitter. MEC applications are usually latency sensitive and computation intensive, such as image recognition in autonomous driving and online gaming. The computation of MEC applications requires execution at the network edge to mitigate long transmission delays between end users and the cloud.

Due to the different deployment architectures, the performance of MEC outweighs that of MCC in terms of latency, energy consumption, context-aware computing, and security and privacy. The benefits of MEC over MCC can be summarized as follows.

- *Latency performance*: The latency of mobile applications is composed of two parts: communication latency and computation latency. Compared with MCC, the propagation distances of MEC systems are much shorter. Generally, the distances of MEC systems are no longer than a kilometer. However, the distances between the cloud center and end users in MCC can be hundreds of kilometers and even span countries or continents. For example, the transmission distances for end users who want to use Google MCC applications in different parts of the world can vary from several kilometers to thousands of kilometers. Moreover, the transmission of MCC data usually requires passage through several networks, including a radio access network and the Internet, which can lead to additional delays in communication for MCC applications. In MEC systems, however, the computation data are transmitted through edge mobile networks or device to device, which are much simpler transmissions than in MCC systems. In terms of computation latency, although the cloud has large amounts of computational resources, they are shared with massive numbers of MCC users. In MEC systems, on the other hand, the computational capabilities of the servers are allocated to limited numbers of end users within their coverage. The gap in available computation capabilities for end users is thus mitigated. With short transmission distances and simple transmission schemes, MEC systems achieve better latency performance than MCC systems. In MEC systems, the latency is usually less than tens of milliseconds, whereas in MCC systems, the latency can be longer than hundreds of milliseconds.
- *Energy consumption*: Energy-consuming computation tasks can be offloaded from end devices to MEC servers, thus reducing the energy consumption of end devices. Specifically, in the Internet of Things (IoT), such offloading prolongs the battery life of IoT devices. In MCC systems, however, the long communication distances of computation data require end devices to maintain high transmission power, which will increase their energy consumption. By offloading computation-intensive tasks to nearby MEC servers, energy consumption is significantly reduced in MEC systems.
- *Context-aware computing*: Since MEC servers are much closer to the end devices, they can interact with end devices in real time by tracking their running states and making instantaneous decisions for them. The real-time interactions between MEC servers and end devices enable users to access context-aware services [1], such as real-time traffic updates and live video feeds, based on users' locations. For example, in autonomous driving, the MEC server leverages the information

from vehicles, such as locations and traffic conditions, to determine the vehicle's driving actions.
- *Security and privacy*: With increasing concerns about data security and privacy, the protection of user data has become a critical issue in mobile applications. With the development of end devices, the data collected can contain much of users' sensitive information. In MCC applications, the user data are transmitted to a remote cloud center over long distances. The data are then managed and processed by the cloud providers, such as Amazon and Microsoft. The risks of data leakage are extremely high during long-distance transmissions and remote management in the cloud. Cloud centers are more prone to become the targets of economically motivated attacks. On the other hand, MEC servers are deployed in distributed architectures of small scale. Many MEC servers can be privately operated and owned by the users in environments such as home cloudlets. Thus the risks of data leakage are considerably mitigated. MEC systems enhance user security and privacy for applications that might need to collect and process private user information.

Although MEC and MCC have different architectures and characteristics, they can sometimes also cooperate together, to enhance the computing capability and latency performance of the system. A series of works have explored combining MCC with MEC to improve application performance. For example, in the application scenario of online gaming, MEC provides users with cached data and local computation, while MCC provides users with new data and intensive computation. Thus the user's experience of image quality and delay performance can be considerably improved.

1.2   Overview of MEC

MEC provides a distributed computing environment by placing compute and storage resources closer to the consumer or enterprise end user. The term *MEC* was first introduced in 2013, when Nokia Siemens Networks and IBM developed a platform called Application Service Platform for Networks to allow mobile operators to deploy, run, and integrate applications at the edge of the network [2]. In 2014, the MEC technical white paper was developed by the European Telecommunications Standards Institute (ETSI) [3], and a new Industry Specification Group was established in ETSI to produce specifications. The Industry Specification Group has delivered several specifications on service scenarios, requirements, architecture, and APIs that will allow for the efficient and seamless integration of applications from vendors, service providers, and third parties across multi-vendor MEC platforms. In 2016, ETSI dropped the word *mobile* from MEC, renaming the technology multi-access edge computing (with the same acronym, *MEC*), to extend its scope to heterogeneous access technologies (e.g., LTE, 5G, Wi-Fi, and fixed access technologies).

MEC is a new paradigm that provides an IT service environment and cloud-computing capabilities at the edge of the network, within the radio access network, and in close proximity to mobile subscribers. The main purpose of MEC is to

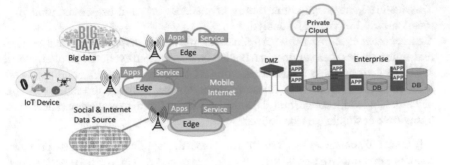

Fig. 1.2 The MEC framework

reduce backhaul network congestion, support low-latency applications, and offer an improved user experience. The general framework of MEC is shown in Fig. 1.2. Different types of big data applications, IoT devices, and social and Internet services are connected to distributed mobile edge networks. The mobile edge networks are connected to the private cloud network via a demilitarized zone for security. The private cloud network is equipped with sufficient databases and applications to provide centralized processing, storage, and computing service. Since cloud services and applications are far from mobile users, MEC deploys the distributed edge services and applications at wireless network infrastructures (i.e., base stations, Wi-Fi access points, or femto access points) to form distributed mobile edge networks. Users can easily access nearby the wireless network infrastructure to enjoy real-time and high-quality service applications. Additionally, MEC not only benefits end users, but also improves resource utility and network efficiency with network optimization, such as computation and caching resource allocation.

According to the ETSI white paper, MEC can be characterized by features such as on premises, proximity, low latency, location awareness, and network context information. These features can be briefly explained as follows.

- *On premises*: MEC platforms can run isolated from the rest of the network while maintaining access to local resources. This is very important for machine-to-machine scenarios, such as security or safety systems that need high levels of resilience.
- *Proximity*: MEC servers are usually deployed close to mobile users. MEC is thus particularly useful in capturing key information for analytics and big data. It is also beneficial for compute-hungry devices, such as augmented reality (AR) and video analytics.
- *Lower latency*: Since MEC services run close to end devices, latency can be considerably reduced, which can be utilized to react faster, improve user experience, or minimize congestion in other parts of the network.
- *Location awareness*: Due to proximity, MEC can leverage signaling information received from edge devices to determine the location of each connected device.

This feature leads to an entire family of business-oriented use cases, including location-based services and analytics.

- *Network context information*: Applications providing network information and real-time network data services can benefit businesses and events by implementing MEC for their business model. Based on real-time radio network conditions and local contextual information, these applications can estimate the radio cell and network bandwidth congestion. This can help in the future to make smart decisions to improve customer service delivery.

MEC not only enhances the performance of existing applications, but also provides tremendous potential for developing a wide range of new and innovative applications. In the following, we introduce several typical use cases in MEC.

- *Internet of Vehicles (IoV)*: The IoV aims to enhance safety, reduce traffic congestion, sense vehicles' behaviors, as well as provide opportunities for numerous vehicular services, such as smart navigation, traffic warnings, and real-time driving route planning. The communication model in IoV can either be vehicle to vehicle (V2V), vehicle to roadside infrastructure (V2R), or vehicle to Internet (V2I). However, resource-constrained vehicles can be strained by computation-intensive applications, resulting in bottlenecks and making it challenging for the vehicles to ensure the required quality of service level. MEC can alleviate the heavy computation requirement of vehicles by providing computation capabilities at the edge of the radio access network [7]. Due to the proximity of MEC servers to vehicles, the offloaded tasks can be accomplished with low latency and high efficiency.
- *Smart grids*: A smart grid offers transparent energy distribution where both consumers and utilities can monitor and control their pricing, production, and consumption in almost real time. A smart grid infrastructure is an electrical grid that consists of several components, such as smart appliances, renewable energy resources, and energy efficiency resources [8]. Smart meters are distributed throughout the network to receive and transmit measurements of energy consumption. All the data collected by the smart meter are supervised by supervisory control and data acquisition systems to maintain and stabilize the power grid. The analysis of the data from various smart meters in the smart grid environment is challenging, since it varies with respect to parameters such as size, volume, velocity, and variety. The usage of MEC can improve performance in throughput, response time, and transmission delay. Distributed smart meters and microgrids, integrated with MEC, have the ability to conduct nearby data management and analysis. For example, a three-tier fog-based smart grid architecture [9] is proposed to extend the capabilities of cloud-based smart grids in terms of latency, privacy, and locality.
- *Unmanned aerial vehicles (UAVs)*: With recent advancements in technology and reductions in manufacturing cost, UAVs have received growing interest in various applications, such as disaster rescue, cargo delivery, filming, as well as monitoring. To maintain UAVs' safe operation with real-time commands and enable the above computation-intensive applications, it is important to enhance the communication and computational capabilities of UAVs. With the help of MEC, edge computing resources can be deployed on UAVs to support computation-intensive and latency-

critical applications. On the other hand, the rapid growth of network traffic has made it difficult for static base stations to support the data demands of billions of devices. UAVs can act as flying base stations to support ubiquitous wireless communications and unprecedented IoT services, due to their high flexibility, easy deployability and operability. In UAV-aided MEC networks, UAVs can act as mobile computation servers or computation offloading routers to provide users better wireless connections and greater flexibility in the implementation of MEC.

- *AR/virtual reality (VR) services*: AR and VR allow users to interact more naturally with virtual worlds based on the data generated from sensory inputs, such as sound, video, graphics, or a global positional system. AR/VR applications need real-time information on users' status, such as their location and direction, and also require low latency as well as intensive data processing for a better user experience. MEC is an ideal solution for AR and VR applications, since MEC servers can exploit local context information and provide nearby data processing. Deploying a VR controller on a MEC server and utilizing wireless links to transmit VR images and audio can increase tracking accuracy, obtaining round-trip latencies of one millisecond and high reliability [4]. Caching parts of VR videos on MEC servers in advance and then performing computations on VR devices can save large amounts of communication bandwidth and fulfill low latency requirements [5]. Offloading computation-intensive tasks to edge servers can increase the computational capacity of AR devices and save their battery life [6].s

- *Video stream analytics*: Video streaming has a wide range of applications, such as vehicular license plate recognition, face recognition, and security surveillance. Video streaming has been observed to comprise more than half of the entire mobile data traffic in current networks, and the percentage is still increasing. The main video streaming operations are object detection and classification. These tasks usually have high computation complexity. If these tasks are processed in the central cloud, the video stream will be transmitted to the cloud network, which will consume a great amount of network bandwidth. MEC can offer ultra-low latency, which is required for live video streaming, by performing the video analysis in a place close to edge devices. A caching-based millimeter-wave framework is proposed to pre-cache content at the base station for hand-off users [11]. The proposed solution can provide consistent high-quality video streaming for high-mobility users with low latency.

1.3   Book Organization

This book aims to provide a comprehensive view of MEC. As a key enabling technology for achieving intelligence in wireless communications, MEC has been widely studied in a series of areas, including edge computing, edge caching, and the IoV. However, with the development of new technologies, such as blockchain, artificial intelligence, and beyond 5G/6G communications, new opportunities have opened up for the fulfillment of MEC and its applications. Motivated by these new changes, this work provides comprehensive discussions on MEC in the new era. We first present the fundamental principles of MCC and MEC technologies. Next, we present applications of MEC in typical edge computing and edge caching scenarios. Furthermore, we discuss research opportunities in MEC in emerging scenarios such as the IoV, 6G, and UAVs. Finally, we provide potential directions of MEC for the future.

This book is organized as follows. Chapter 2 presents the models and policies of edge computing. Chapter 3 describes the architecture and performance metrics of mobile edge caching. A case study of deep reinforcement learning–empowered edge caching is further conducted in Chap. 4. Applications of MEC in the IoV for task and computation offloading are presented in Chap. 5. Chapter 6 describes details on the application of MEC to UAVs. Finally, Chap. 7 provides a comprehensive discussion of the future of MEC.

Chapter 2
Mobile Edge Computing

Abstract Mobile edge computing is a promising paradigm that brings computing resources to mobile users at the network edge, allowing computing-intensive and delay-sensitive applications to be quickly processed by edge servers to satisfy the requirements of mobile users. In this chapter, we first introduce a hierarchical architecture of mobile edge computing that consists of a cloud plane, an edge plane, and a user plane. We then introduce three typical computation offloading decisions. Finally, we review state-of-the-art works on computation offloading and present the use case of joint computation offloading.

2.1 A Hierarchical Architecture of Mobile Edge Computing (MEC)

To better understand the internal logic of MEC, we first present a hierarchical architecture that vertically divides the edge computing system into three layers: the user layer, the edge layer, and the cloud layer, as shown in Fig. 2.1. The user layer is distinguished by the wireless communication mode between mobile devices and wireless infrastructures. The edge and cloud layers mainly refer to the computing resources of the edge and cloud servers, respectively.

Devices in the user layer include sensors, smartphones, vehicles, smart meters, and radio-frequency identification devices. These devices access edge servers via wireless communication and then offload computation-intensive tasks to the lightweight, distributed edge servers to process. According to wireless network topology and communication modes, the communication between mobile devices and a wireless infrastructure can be split into the following three modes.

- Heterogeneous network: Next generation wireless networks will run applications that require large demand for high data rates. One solution to help reduce the data rate requirement is the densification of the network by deploying small cells. Such densification results in higher spectral efficiency and can reduce the power consumption of a mobile device due to its communication with small nearby cell base stations. This solution significantly improves network coverage. The concurrent operation of macro base stations (MBSs) and micro, pico, femto, and

© The Author(s) 2022
Y. Zhang, *Mobile Edge Computing*, Simula SpringerBriefs on Computing 9,
https://doi.org/10.1007/978-3-030-83944-4_2

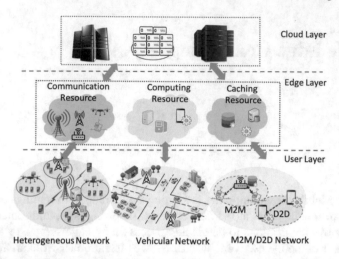

Fig. 2.1 Hierarchical MEC architecture

unmanned aerial vehicle–aided base stations is termed a heterogeneous network. In heterogeneous networks, all base stations are equipped with computational resources and artificial intelligence functions. Resource-limited mobile devices can offload their tasks to these heterogeneous base stations, which can then utilize a fine-grained computational resource allocation policy to process the offloaded tasks.

- Vehicular network: Vehicular networks are inseparable from a smart city environment, owing to several applications that improve the quality of life, safety, and security. A vehicular network is formed among moving vehicles, roadside units, and pedestrians, which can be deployed in rural, urban, and highway environments. Vehicle-to-everything communication allows vehicles to communicate with other vehicles and their surroundings via wireless links. Vehicle-to-everything communication has three main scenarios: vehicle to vehicle, vehicle to infrastructure, and vehicle to pedestrian [12]. Commonly used technologies are dedicated short-range communications, IEEE 802.11p, the IEEE 1609 family of standards, and Long Term Evolution (LTE). With advancements in communication technologies, a number of promising applications are emerging for vehicular networks. These vary from safety applications, such as blind spot warning and traffic light violations to entertainment, such as streaming media, or convenience, such as parking space identification. In vehicular networks, ubiquitous edge resources can be deployed on nearby infrastructures to offer vehicles a high quality of service. Compared to common mobile nodes, vehicles can move at quite high speeds, which causes the topology of a vehicular network to frequently change. Detailed policy design must carefully consider such dynamic network topologies.

- Mobile-to-mobile (M2M)/device-to-device (D2D) networks: M2M is an enabling technology for the Internet of Things, which involves autonomous connectivity and communication among devices from embedded sensors and actuators to powerful computationally rich devices without human intervention. D2D allows devices to communicate with each other through a direct wireless link without traversing the base station or core network. With the technological advancement of smart devices, more computing and caching resources are distributed among the end users. Computational tasks can thus be offloaded not only to edge servers, but also to devices in D2D and M2M networks.

The edge layer is located in the middle of the hierarchical architecture and consists of multiple distributed edge servers to provide distributed intelligent wireless computing for users. Edge servers can be deployed in the network infrastructure, such as base stations, roadside units, wireless access points, and gateways, or they can be mobile phones, vehicles, tablets, and other devices with computing and storage capabilities. Generally, edge servers are widely distributed in hotspots such as cafes, shopping centers, bus terminals, streets, and parks. Given the proximity of edge servers to end users, computing-intensive and delay-sensitive tasks can be offloaded and accomplished with low latency and high efficiency. There are three types of resources in the edge layer: communication resources, caching resources, and computing resources. Communication resources refer to bandwidth, spectrum, and transmission power. Computing resources mainly refer to CPU cycles. Caching resources are related to the memory capacity on edge servers. Since edge servers are ubiquitously distributed, their computing and caching resources capacities are usually limited. The full use of edge resources requires the joint optimization of communication, caching, and computing resources.

The central cloud layer consists of multiple servers with strong processing, caching, and computing capabilities. With a global view, this layer can leverage advanced techniques such as data mining and big data, for a network-level orchestration shift from reactive to proactive network operation, by predicting events or pre-allocating resources. With their high computing capability and sufficient caching resources, cloud servers can process delay-tolerant applications and store larger or less popular content. Further, the central cloud layer can effectively manage and control multiple edge servers and provide them with secure connections.

2.2   Computation Model

Computation offloading is an approach to offload computation-intensive and delay-sensitive tasks to resource-rich edge servers and/or cloud servers to process. This approach can help prolong the battery life of mobile devices and reduce task processing latency. The key problems in computation offloading are in deciding whether to offload, the amount of the computation task that needs offloading, and which server to offload to. Basically, computation offloading can result in the following three types of decisions [13], as shown in Fig. 2.2:

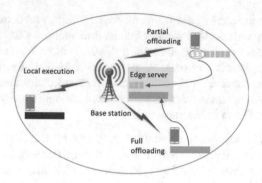

Fig. 2.2 Computation offloading decision

- Local execution: The entire computation task is completed locally. If the computational resources of the edge servers are unavailable or the wireless channel is of
  poor quality, which can result in high transmission latency, local execution can be
  preferred.
- Full offloading: The entire computation task is offloaded and processed by an edge
  server.
- Partial offloading: Part of the computation task is processed locally while the rest
  is offloaded to an edge server.

The computation offloading decision is very difficult, since it requires considering
multiple factors, such as application requirements, the quality of the communication
link, and the computing resource capacities of edge servers.

2.2.1 Computation Model of Local Execution

As the noted above, the CPU is the primary engine for computation. The CPU's
performance is controlled by CPU cycles f_m. The state-of-the-art mobile CPU architecture adopts an advanced dynamic frequency and voltage scaling technique, which
allows for stepping up or down CPU cycles, increasing and reducing energy consumption, respectively. In practice, the value of f_m is bounded by a maximum value,
f_{max}, which reflects the limitation of the mobile's computation capability. A computation task can be described as $D \triangleq (d, c, T)$, where d denotes the data size of the
computation task, c is the required number of CPU cycles for computing one bit of
the computation task, and T denotes the maximum latency allowed to accomplish
the task. The local execution time for a computing task D can now be expressed as

$$T^L = \frac{dc}{f_m} \tag{2.1}$$

which indicates that more CPU cycles are required to reduce the execution latency.

Since devices are energy constrained, the energy consumption of local execution is a critical performance metric for computing efficiency. According to [14], the energy consumption of each CPU cycle is given by ςf_m^2, where ς is the effective switched capacitance, depending on the chip architecture. The energy consumption for executing task D with f_m CPU cycles can be derived as

$$E^L = \varsigma dc f_m^2 \tag{2.2}$$

From (2.1) and (2.2), if T^L is greater than the maximum latency or if the device's battery capacity is less than E^L, the device should offload the task to edge servers to process. Otherwise, local execution can support the computation task.

2.2.2 Computation Model of Full Offloading

In this section, we present two computation models of the full offloading for a single-user MEC system and a multi-user MEC system, respectively.

The single-user MEC system is the simplest case and consists of a single device and a single edge server. Denote F_e as the computational resource capacity of the edge server. The device offloads the entire computation task to the edge server to process. The task computation time is thus given by

$$t^{F,computing} = \frac{dc}{F_e} \tag{2.3}$$

Since the process of offloading involves wireless transmission, the total task execution time is the sum of the task computation time and the task transmission time, which can be expressed as

$$T^{F,s} = \frac{dc}{F_e} + \frac{d}{r^s} \tag{2.4}$$

where r^s is the wireless transmission data rate between the device and the edge server. The energy consumption for completing the offloaded computation task also includes two parts: the energy consumption for computation and the energy consumption for wireless transmission. The total energy consumption can be expressed as

$$E^{F,s} = \varsigma dc F_e^2 + p\frac{d}{r^s} \tag{2.5}$$

where p is the transmission power of the device.

In the multi-user MEC system, several devices can be associated with the same edge server and offload their tasks to the edge server simultaneously. In this case, each device is assigned only to a part of the edge server's computational resources. Denote the computation task of device i as $D_i \triangleq (d_i, c_i, T_i)$, where d_i denotes the data size of the computation task on device i, c_i is the required number of CPU cycles for computing one bit of the computation task, and T_i denotes the maximum latency allowed to accomplish the task. Let f_e^i be the computational resources that the edge server allocates to device i. Since the process of offloading involves wireless transmission, the total task execution time of device i can be expressed as

$$T_i^{F,m} = \frac{d_i c_i}{f_e^i} + \frac{d_i}{r_i^m} \tag{2.6}$$

where r_i^m is the wireless transmission data rate between device i and the edge server.

The corresponding energy consumption of completing the offloaded computation task of device i can be expressed as

$$E_i^{F,m} = \varsigma d_i c_i (f_e^i)^2 + p_i \frac{d_i}{r_i^m} \tag{2.7}$$

where p_i is the transmission power of device i.

Different from the single-user MEC system, devices in the multi-user MEC system share the same computational resources and wireless channel. Therefore, computational resource allocation, channel assignment, bandwidth allocation, and power control should be jointly optimized. Since the total computational resources of the edge server are limited, there is a computational resource constraint (i.e., $\sum_i f_e^i \leq F_e$). A more complex model considering a multi-user multi-MEC server was proposed in [14, 15]. With the dense deployment of MEC servers, a joint user association and computation offloading scheme was designed in [14], and a joint communication resource allocation and computation offloading scheme was designed in [15].

2.2.3 A Computation Model for Partial Offloading

Partial offloading is a very complex process that can be affected by different factors, such as the offloadability of an application [16], the dependency of the offloadable parts [17], and user preferences and channel connection quality [32]. To simplify the description, we assume each computation task can be offloaded and arbitrarily divided into two parts. One part is executed on the device and the other is offloaded to an edge server for edge execution.

Let λ ($0 \leq \lambda \leq 1$) be the offloading ratio, which represents the ratio of the offloaded task to the total task. That is, an amount λd is offloaded to the edge server to be computed and the rest, $(1 - \lambda) d$, is computed locally. The task computation time upon partial offloading can be expressed as

$$t^{P,computing} = \frac{(1-\lambda)dc}{f_m} + \frac{\lambda dc}{F_e} \tag{2.8}$$

Since one part of the computation task (i.e., λd) involves wireless transmission, the total time for completing this task can be expressed as

$$T^P = \frac{(1 - \lambda)dc}{f_m} + \frac{\lambda dc}{F_e} + \frac{\lambda d}{r} \tag{2.9}$$

The energy consumption required for completing this task consists of three parts:

$$E^P = \varsigma(1 - \lambda)dcf_m^2 + \varsigma\lambda dcF_e^2 + p\frac{\lambda d}{r} \tag{2.10}$$

where the first term indicates the local energy consumption for processing the amount $(1 - \lambda)\,d$, the second term indicates the energy consumption for processing the amount λd on the edge server, and the third term is the energy consumption of the wireless transmission. In partial offloading, the key problem is to decide the offloading ratio, considering system constraints. For example, if the energy or computational resources of the device are almost used up, offloading the task to the edge server is desirable (i.e., the offloading ratio should be close to one). If the quality of the wireless channel is poor or the available computational resources of the edge server are limited, local execution could be a better choice. Note that the above models can be easily extended to the multi-user MEC system.

2.3 Offloading Policy

The key problem in edge computing is making the offloading decision. According to the previous description, the results of the offloading decision are either local execution, full offloading, or partial offloading. Combining local execution and full offloading, the problem can be modeled as a zero–one binary offloading problem. Partial offloading can be modeled as a continuous offloading decision making problem. First, we introduce the research on binary offloading in the next section.

2.3.1 Binary Offloading

Binary offloading mainly involves small-scale computation tasks that have high computational resource requirements. Such tasks will be offloaded in entirety to the edge server. Computing offloading can effectively reduce the task completion delay and save the energy consumption of devices. When the device does not choose offloading (i.e., local execution), the task completion delay involves only the local task computation time. When the device chooses offloading, the task completion delay involves three parts: (1) the wireless transmission time of the computation task from the device to the edge server, (2) the task computation time spent on the edge server, and (3) the wireless transmission time of the computation result from the edge server to the device. Similarly, when the device does not offload the task, the total energy consumption required to complete the task includes only local task computation energy consumption. If the device offloads any of the computation task, the total energy consumption consists of two parts: the energy consumption of the wireless transmission from the device to the edge server and the energy consumption of the computation on the edge server.

2.3.1.1 Minimization of Task Execution Delay

The authors in [18] proposed a one-dimensional search algorithm to minimize execution delay. The proposed algorithm can find an optimal offloading decision policy based on the buffer state, available processing power, and channel information. The offloading decision determines whether to process the application locally or at the MEC server. Another idea aimed at minimizing the execution delay was introduced in [20]. Compared to [18], these authors considered users applying dynamic voltage and frequency scaling and proposed a low-complexity Lyapunov optimization-based dynamic computation offloading algorithm. This algorithm allows users to make an offloading decision in each time slot and simultaneously allocates CPU cycles and transmission power. The proposed method can reduce execution times by up to 64% by offloading the computation task to the edge server. Different from the two works focusing on the design of computation offloading algorithms, the authors in [19] proposed an MEC-assisted offloading architecture that allows for deploying intelligent scheduling logic, namely, a mobile edge scheduler, at the MEC without requiring large computational resources at the eNodeB hardware. The introduced mobile edge scheduler runs on the eNodeB. A two-stage scheduling process was proposed to minimize the delay of general traffic flows in the LTE downlink via the MEC server deployed at the eNodeB.

2.3.1.2 Minimization of Energy Consumption

The computation offloading decision to minimize the energy consumption of devices was proposed in [21]. These authors formulated the optimization problem as a constrained Markov decision process. To solve the optimization problem, two types of resource allocation strategies accounting for both computational and radio resources were introduced. The first strategy is based on online learning, where the network adapts dynamically with respect to the application running on the device. The second, precalculated offline strategy is based on prior knowledge of the application properties and statistical behavior of the radio environment. Numerical experiments showed that the precalculated offline strategy can outperform the online strategy by up to 50% for low and medium arrival rates (loads). Since the offline strategy proposed in [21] showed its merits, the authors in [22] proposed two additional offline dynamic programming approaches to minimize the average energy consumption of devices.

One of the dynamic programming approaches to find the optimal radio scheduling offloading policy is deterministic, while the other is randomized. Numerical experiments showed both offline policies can reduce energy consumption compared to offloading-only and static processing strategies. The authors in [22] further extended the work in [23] from single user to multi-user by jointly optimizing resource allocation and computation offloading to guarantee fairness between users, low energy consumption, and average queuing/delay constraints. Another multi-user offloading decision strategy was proposed in [24] to minimize system energy consumption. This paper determined three multi-user types based on the time and energy cost of the task computing process. The first type of user can compute tasks on the MEC server. The second type of user computes the task on local equipment. The third type of user can decide to either implement tasks locally or offload tasks to the MEC server. Based on the user classification, a joint computation offloading and radio resource allocation algorithm was proposed. The proposed algorithm can decrease energy consumption by up to 15% compared to computation without offloading.

2.3.1.3   Trade-Off Between Energy Consumption and Execution Delay

A computation offloading decision for a multi-user multi-task scenario was proposed in [25] to make the trade-off between energy consumption and execution delay. These authors considered jointly the offloading decisions for all the tasks of each user and the sharing of computational and communication resources among all the users as they compete to offload tasks through a wireless link with limited capacity. The computation offloading problem is formulated as a non-convex quadratically constrained quadratic program. To solve this problem, an efficient three-step algorithm was designed that involves semidefinite relaxation, alternating optimization, and sequential tuning. The numerical results showed the proposed algorithm outperformed purely local processing, purely cloud processing, and hybrid local–cloud processing without an edge server. Another algorithm for the computation offloading decision to trade off energy consumption and execution delay was proposed in [26]. The main difference between the works [25, 26] is that the task in [25] can be also offloaded to a remote centralized cloud if the computational resources of the MEC are insufficient. In [26], the authors proposed a computation offloading decision to minimize both the total task execution latency and the total energy consumption of mobile devices. Two cases of mobile devices were considered: devices with a fixed CPU frequency and those with an elastic CPU frequency. In the fixed CPU scenario, a linear programming relaxation–based algorithm was proposed to determine the optimal task allocation decision. In the elastic CPU scenario, the authors first considered an exhaustive search–based algorithm and then utilized a semidefinite relaxation algorithm to find the near-optimal solution.

2.3.2   Partial Offloading

The literature cited above focused on binary offloading strategies. In a binary offloading problem, the computing task is considered as a whole. However, in practical applications, computing tasks are often divided into multiple parts [27]. According to the divisible nature of computing tasks, devices can offload part of a task, rather than its entirety, to the edge server. There are thus two types of tasks: (1) tasks that can be divided into multiple discrete segments that can all be offloaded to the MEC server for execution and (2) tasks that can be split into two consecutive parts, non-offloadable and offloadable, and only the offloadable part can be offloaded. Next, we introduce works focused on partial offloading.

2.3.2.1   Minimization of Task Execution Delay

The authors in [28] investigated a latency minimization resource allocation problem for a multi-user offloading system with partial offloading. A partial compression offloading was proposed that has three steps. First, each device compresses part of the raw data locally and then transmits the compressed data to the edge server. Second, the device transmits the remaining part of the raw data to the edge server, which compresses the data. Finally, the edge server combines the two parts of compressed data in the cloud center. A weighted sum latency minimization partial compression offloading problem was formulated and an optimal resource allocation algorithm based on the subgradient was designed. More general work on partial offloading was covered in [29]. The authors jointly considered a partial offloading and resource allocation scheme to minimize the total latency for a multi-user offloading system based on orthogonal frequency division multiple access. The proposed scheme first determines the optimal offloading fraction to ensure that the edge computing delay is less than the local execution delay. Then, the proposed scheme determines how to allocate the communication and computational resources. Additionally, users can make full use of multi-channel transmissions to further reduce the transmission delay for tasks with a large data size. The simulation results show that the proposed scheme achieves 17% and 25% better performance than random and complete offloading schemes, respectively.

2.3.2.2   Minimization of Energy Consumption

In [27], the authors investigated partial computation offloading to minimize the energy consumption of devices by jointly optimizing the CPU cycle frequency, the transmission power, and the offloading ratio. They designed an energy-optimal partial computation offloading algorithm that transformed the non-convex energy consumption minimization problem into a convex one based on the variable substitution technique and obtained a globally optimal solution. The authors also analyzed the conditions under which local execution is optimal. Analyzing the optimality of total offloading, the authors concluded that total offloading cannot be optimal under dynamic voltage scaling of the device. The authors in [30] proposed a joint scheduling and computation offloading algorithm for multi-component applications using an integer programming approach. The optimal offloading decision involves which components need to be offloaded, as well as their scheduling order. The proposed algorithm provides a greater degree of freedom in the solution by moving away from a compiler predetermined scheduling order for the components toward a more wireless-aware scheduling order. For some component dependency graph structures, the proposed algorithm can shorten execution times by the parallel processing of appropriate components on the devices and in the cloud. To minimize the expected energy consumption of the mobile device, an energy-efficient scheduling policy for collaborative task execution between the mobile device and a cloud clone was proposed in [31]. The authors formulated the energy-efficient task scheduling problem as a constrained stochastic shortest path problem on a directed acyclic graph. They also considered three alternative stochastic wireless channel models: the block fading channel, the independent and identically distributed stochastic channel, and the Markovian stochastic channel. To solve the formulated problem, the authors leveraged a one-climb policy and designed a heuristic algorithm to determine the task execution decision.

2.3.2.3 Trade-Off Between Energy Consumption and Execution Delay

Partial offloading decision considering a trade-off between energy consumption and execution delay was described in [32]. The offloading decision considered four parameters: (1) the total number of bits to be processed, (2) the CPU cycles of the device and of the MEC server, (3) the channel state between the device and the serving femtocell access points, and (4) the device's energy consumption. The joint communication and computational resource allocation problem was formulated as a convex optimization problem. The simulation results indicated that partial offloading could reduce the energy consumption of devices, compared to the case of full offloading, when all the computation tasks are forced to be carried out on either the device or at the femtocell access point. The study in [33] provided a more in-depth theoretical analysis on the trade-off between energy consumption and the latency of the offloaded applications preliminarily handled in [32]. To carry out partial offloading, the authors considered data partition-oriented applications and focused on three parameters of an application: (1) the size of the data, (2) the completion deadline, and (3) the output data size. Then, a joint optimization of the radio and computational resource problem was formulated, and a simple one-dimensional convex numerical optimization technique was utilized to solve it. The authors further demonstrated that the probability of computation offloading is higher when given good channel quality. The authors in [34] considered the trade-off between power consumption and execution delay for a multi-user scenario. The authors formulated a power consumption minimization problem with an application buffer stability constraint. An online algorithm based on Lyapunov optimization was proposed that decides the optimal CPU frequency of the device for local execution and allocates the transmission power and bandwidth when offloading the application to an edge server. The numerical results demonstrated that computation offloading can reduce power consumption up to roughly 90% and reduce execution delays by approximately by 98%.

2.4 Challenges and Future Directions

A wide variety of research challenges and opportunities exists for future research on computation offloading. However, the MEC research is still in its infancy, and many critical factors have been overlooked for simplicity. In this section, we point out several open challenges and shed light on possible future research directions.

- *Multi-server scheduling*: The collaboration of multiple MEC servers allows for their resources to be jointly managed in serving a large number of mobile devices simultaneously. Server cooperation not only can improve resource utilization but also can provide mobile users with more resources to enhance user experience. However, the increase in network size hinders practical MEC server scheduling. Too many offloading users will cause severe inter-user communication interference and the system will need to make large numbers of offloading decisions. More comprehensive research is required for multi-server scheduling.
- *Multi-resource optimization*: The architecture of mobile edge networks involves various resources: computing, caching, and communication resources. The efficient integration of these resources to achieve optimal performance for all users and applications is quite challenging. Efficient resource management requires the design of distributed low-complexity resource optimization algorithms, considering radio and computational resource constraints and computation overhead.
- *User mobility*: User mobility is a key challenge in mobile edge networks. Since the movement and trajectory of users provide location and personal preference information for edge servers, the contact times between users and MEC servers is dynamic, which will impact the offloading strategy. Moreover, the frequent mobility of users causes frequent handovers among edge servers, which will increase computation latency and thus deteriorate user experience. Therefore, mobility management techniques from both horizontal and vertical perspectives should be implemented to allow users seamless access to edge servers.
- *Security*: Security is one of the main concerns of technology advisers in securing MEC deployments. The deployment of edge cloud servers is creating novel security challenges due to the exploitation of mobile device information. The growing rate of the evolution of security solutions cannot keep up with the pace of new security challenges. Many existing security protocols assume full connectivity, which is not realistic in mobile edge networks, since many links are intermittent by default. On the other hand, in MEC, user data are offloaded to an MEC server

that gives access control to other mobile users. This introduces challenges, such as data integrity and authorization. For example, offloaded data can be modified or accessed by malicious users. Moreover, data owners and data servers possess dissimilar identities and business interests that make the scenario more vulnerable. Therefore, a comprehensive scientistc research study is required to avoid any security issues that can damage MEC systems.

This chapter first introduced the hierarchical mobile edge computing architecture with a cloud plane, an edge plane, and a user plane. Then, three types of computation models were discussed in detail for the typical computation offloading problem in MEC. In terms of the offloading decision, current research on computation offloading was surveyed, as were the binary offloading and partial offloading problems. Finally, several open challenges and future directions were discussed.

Chapter 3
Mobile Edge Caching

Abstract Edge caching is a key part of mobile edge computing. It not only can support the necessary task data for edge computing, but also enables powerful Internet of Things applications with massive amounts of data and various types of information in access networks. In this chapter, we present the architecture of the edge caching mechanism and introduce metrics for evaluating caching performance. We then discuss key issues in caching topology design, caching data scheduling, as well as caching server cooperation and present a case study of artificial intelligence–empowered edge caching.

3.1 Introduction

In recent years, wireless communication has witnessed an explosive growth of smart Internet of Things (IoT) devices and powerful mobile applications that greatly facilitate our daily life and improve manufacturing efficiency. The implementation of these applications requires massive content input and relies heavily on high-speed, low latency data transmission. However, the backhaul links between content servers and wireless access points are always constrained by bandwidth, making the stringent transmission requirements hard to meet. This data hungry characteristic poses significant challenges for mobile networks, especially in application scenarios with a large scale of smart devices. Mobile edge caching is a promising approach to alleviate backhaul load and address these challenges. It utilizes caching resources at the edge nodes and provides popular content access close to the end users.

3.2 The Architecture of Mobile Edge Caching

To clearly illustrate the components and operation mechanism of edge cache systems, we present a hierarchical mobile edge caching architecture, which is illustrated in Fig. 3.1. This architecture consists of four layers, namely, an application layer, a user node layer, an edge server layer, and a cloud service layer.

© The Author(s) 2022
Y. Zhang, *Mobile Edge Computing*, Simula SpringerBriefs on Computing 9,
https://doi.org/10.1007/978-3-030-83944-4_3

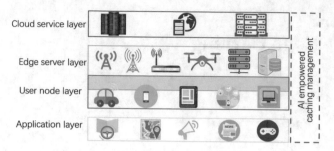

Fig. 3.1 Hierarchical mobile edge caching architecture

The application layer contains a variety of wireless mobile applications with intensive data input requirements. For instance, the driving control of autonomous vehicles depends on high-resolution map data. Furthermore, the map data provided to the vehicles should be updated in real time as traveling vehicles reach different areas. Another example is the daily news broadcast. Although the data delivery does not have strict delay constraints, since the proportion of multimedia news continues to increase, a large amount of the media data will pose a heavy transmission burden on wireless networks. To distinguish different applications according to their caching service demands, we model the data caching task as (f, t^{max}), where f is the amount of data required and t^{max} denotes the maximum latency tolerance before the data are received.

Above the application layer is the user node layer, which is composed of multiple forms and types of devices and physical entities, including mobile phones, wearable devices, tablets, and smart vehicles. It is worth noting that the user nodes can be data requesters while also acting as data providers. A typical example is a smartphone. When a video that has not been downloaded onto the phone is played, the phone can ask for data from a remote video server. During the video playback, the phone can provide the cached data to other phones in device-to-device (D2D) mode. Consequently, parts of user nodes can be regarded as special data servers and classified at the edge server level. The characteristics of the user nodes can affect the data caching performance, including the node's speed of motion, wireless transmission power, and communication topology.

The edge server layer focuses on providing data to user nodes in proximate areas at low cost. To do so, edge caching servers are usually installed in cellular network base stations, roadside units (RSUs), Wi-Fi access points, unmanned aerial vehicles, and other access network infrastructures. On the one hand, cached data can be delivered by directly using the wireless transmission capabilities of these infrastructures; on the other hand, the power supply facilities of these infrastructures can be utilized to provide adequate electricity to the edge servers. These servers are associated with two key performance evaluation indicators. One is the data storage capacity, and the other is the service coverage radius, which is determined by the transmission power of the attached wireless infrastructures and their operating environments.

When edge servers cannot meet data demands due to limitations in cache capacity, the caching nodes in the cloud service layer will provide users with a supplementary backup. These cloud servers are powerful and usually have vast amounts of storage space and can thus cache large amounts of data. Although the cloud server can be located as far away from the user nodes as an application server, there are essential differences between the two. As data generators, application servers are scattered throughout various networks. Access to their data can be severely restricted by their communication capabilities. In contrast, cloud servers are mostly located in the core network, equipped with high-speed input–output equipment and high-capacity transmission facilities, with, for example, optical fiber, facilitating high-speed data access for users.

The operation of edge caching requires efficient management, so a caching control module is introduced into the system that is responsible for tracking data popularity and scheduling storage resources. Considering that the caching scheduling relies on the collaboration of various types of entities, including both the data servers and the requesters, this module is implemented across multiple layers. With the advancement of artificial intelligence, powerful machine learning approaches have been implemented within the control module. They help extract data features in complex mobile application environments, predict data demand trends, and fully tap the data caching potential of large-scale heterogeneous service devices.

3.3 Caching Performance Metrics

Many studies are devoted to edge cache optimization, adopting various technical approaches to address different problems in server layout, resource scheduling, data management, and so forth. To quantitatively evaluate the pros and cons of these approaches, the following caching performance metrics are introduced.

3.3.1 Hit Rate Ratio

The hit rate ratio is one of the most important performance metrics in edge caching. To understand this metric, we first need to define the concept of a hit. In the edge caching process, if a user can directly obtain the requested content from an edge cache server, without the help of cloud service facilities or remote data sources, this data access is called a hit. Thus, the hit rate ratio can be calculated by dividing the number of cache hits by the total number of cache hits and misses, as follows:

$$r = \frac{h}{h + m} \tag{3.1}$$

where h is the number of cache hits and m is the number of cache misses. It is worth noting that m contains not only content obtained from the cloud or remote servers, but also content that was not successfully obtained due to transmission failure.

The hit rate ratio reflects how effective a cache is at fulfilling requests for content; in other words, it measures the utilization efficiency of edge service resources in satisfying end users. To improve the performance metric, content popularity–based caching strategies are usually adopted to determine the type of cached data and data update frequency.

3.3.2 Content Acquisition Latency

Content acquisition latency is another chief measurement in caching services that indicates the total time cost, from the user generating a data request to obtaining the complete data. Content acquisition latency can be formally represented as

$$T_a = t_{\text{req}} + \sum_{k \in \mathcal{K}} t_{\text{res},k} \tag{3.2}$$

where t_{req} is the time for a request to be transmitted from a user to the chosen server. Since the request message is small, t_{req} can usually be ignored. If the cache server directly requested by the user does not have the required data, it needs to obtain the resource from a remote server that is $\mathcal{K}$ hops away. The variable $t_{\text{res},k}$ is the one-hop transmission delay of the data response.

Many mobile applications are now delay sensitive. For instance, in mobile navigation, digital map data need to be delivered to the user's smartphone for route planning when the user first enters an area. To improve the timeliness of news, breaking news needs to be pushed to the user end as soon as possible.

The latency metric can be used in two ways in caching service management: one way is to minimize it as an optimization objective, and the other is to make it a constraint of an optimization problem. The content acquisition delay is affected by the location of the data source server and the transmission rate. Approaches to reduce latency can therefore be considered in three aspects: the effective placement of edge caching servers, the timely update of stored data, and the optimized scheduling of communication resources.

3.3.3 Quality of Experience (QoE)

The QoE is a metric that evaluates the performance of edge caching services from the perspective of data users. Since multiple users can have diverse data requirements, QoE can be reflected in different aspects. When a user downloads a large video through cellular networks, the user might mainly concern about the data transmission cost. For content that is urgently needed, users will focus on the timeliness of the data acquisition. A user receiving an advertisement push will care about whether the content meets his or her shopping demands. Considering that users can have multiple preferences at the same time, QoE can also be formed as a combination of multiple factors, as follows:

$$QoE = \sum_{n=1}^{N} \alpha_n \cdot q_n \tag{3.3}$$

where q_n is the nth type of metric value in the QoE evaluation, and α_n is a coefficient that reflects the user's subjective measurement of the importance of different metrics. Since a positive data acquisition experience will encourage users to spend more on edge caching services, which will increase the revenue potential of the edge server operator, QoE has become an important criterion for cache design and management.

3.3.4 Caching System Utility

Unlike QoE, which only considers user-side gains, a system utility metric introduces service-side revenue factors into the caching performance evaluation. Edge cache servers are usually deployed and managed by operators, and the servers' operation process produces energy, maintenance, and other costs. On the other hand, operators can obtain rewards by providing data services to end users. Rational operators aim to reduce costs while maximizing profits. From the perspective of the service side, caching utility can be defined as the difference between the profits and costs. Furthermore, since an edge caching system includes both users and services, we define the system utility as the sum of the user- and service-side utilities, which can be presented as

$$U_{sys} = R_{server} - C_{server} + G_{user} - C_{user} \tag{3.4}$$

where R_{server} and C_{server} are the reward and operation costs of the caching servers, respectively; G_{user} denotes the user utility gained from the edge caching service, such as improved QoE or savings in remote transmission consumption; and C_{user} is the cost paid by the users to access the service, for example, the price paid to the server operator. Generally, the value of C_{user} is equal to R_{server} in a given edge caching system. To improve the caching system utility, we can adopt a game-theoretic approach and form a cooperative game between the users and servers that helps find caching scheduling strategies that benefit both sides.

3.4   Caching Service Design and Data Scheduling Mechanisms

Edge caching is expected to provide data storage services for users to access low latency content, while releasing the burden on backhaul networks between the data generator and end users. Thanks to the evolution of wireless communication and the development of IoT technology, pervasive smart nodes can interact with each other and become integrated so that edge cache services can be implemented on diverse types of nodes and in various network segments. In this section, we investigate an edge cache design from the perspective of how the edge caching service is deployed at heterogeneous nodes, and we then discuss the corresponding caching data scheduling mechanisms. Without loss of generality, we divide the nodes by infrastructures and user devices and correspondingly classify the cache service modes into three types, namely, infrastructure supported, user device sharing enabled, and a hybrid type, as illustrated in Fig. 3.2.

3.4.1   Edge Caching Based on Network Infrastructure Services

The network infrastructure is a vital part of a communication system, and it comprises hardware and software that enables communication connectivity between user devices and service facilities. In this section, infrastructure refers specifically to the base stations of cellular networks, the RSUs in vehicular networks, and the wireless access points in Wi-Fi networks, which are always managed by network operators.

The above-mentioned infrastructures have common characteristics. First, they have a large data cache space that usually reaches several megabytes or even gigabytes. Second, the infrastructures serve a wide coverage area, delivering cached data to multiple user nodes. Through the coordination of multiple infrastructure nodes to form a mesh network, their service range can be further improved. Finally, the infrastructures have fixed locations, and it is hard to spatially adjust the caching service pattern of an individual infrastructure node.

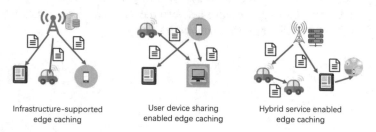

| Infrastructure-supported edge caching | User device sharing enabled edge caching | Hybrid service enabled edge caching |

Fig. 3.2   Edge caching service modes

In view of the above characteristics, infrastructure-based edge caching mechanisms mainly focus on data popularity determination, target server selection, and cache data updates. It is worth noting that, although infrastructures cannot change their locations, they can coordinate with multiple servers distributed in different areas and adjusting their caching strategies to meet the dynamic spatial distribution of data demands. The caching strategy of infrastructure-based servers can generally be described as $a_{m,n}^t = \{0, 1\}$, where m and n are the server node index and the type of data, respectively, and t denotes the time slot for the caching strategy. Here $a_{m,n}^t = 1$ indicates the data are cached, and vice versa.

A few previous works have examined infrastructure-based edge caching. Xu, Tao, and Shen [35] used small base stations as edge caching servers and investigated cache placement optimization to minimize the system's long-term transmission delay without knowledge of user data preferences. The optimization scheme was formulated as a sequential multi-agent decision problem. Moreover, the authors proposed a direct online learning cache strategy in both stationary and non-stationary environments that achieves a good balance between gains in base station coordination and computational complexity. Wang et al. [36] presented an edge caching scenario for mobile video streaming in which base stations distributed citywide provide video storage capacity. Based on the analysis of viewers' request patterns behind both spatial and temporal dimensions, the authors proposed a multi-agent deep reinforcement learning–empowered caching scheme that minimizes both content access latency and traffic costs.

3.4.2 Edge Caching Based on D2D Services

Although infrastructure-based edge caching has produced a promising paradigm to push data closer to the network edge with low transmission latency, there are still problems to be addressed. For instance, the data delivery of infrastructures usually has a large wireless coverage range and works in broadcast mode. Excessive transmission distances can weaken the space-division multiplexing of the spectrum, thereby reducing the efficiency of data caching services. Furthermore, multiple user nodes can have different types of data demands. However, it is hard for the broadcast delivery mode to provide edge services with demand recognition and differentiation.

To address these problems, we resort to D2D caching services. Driven by the advancement of IoT technology, smart devices are becoming increasingly popular. These devices are equipped with a certain data caching capacity and multiple types of wireless communication interfaces, such as cellular, Wi-Fi, and Bluetooth, making the devices potential data carriers and deliverers.

Technical challenges, however, arise in using smart devices to provide efficient caching services. One of the challenges is the dynamic network topology caused by device movement, which makes the stable maintenance of data services for given areas very difficult. In addition, multiple device pairs could concurrently communicate and deliver cached data. It is not easy to efficiently schedule communications in a distributed scenario. Last but not least, constrained by its limited cache resources, an individual user's device cannot provide cache services for large files.

In response to the above challenges, academics have carried out in-depth research. Some works have focused on the analysis of device mobility patterns and leveraged the mobility to expand caching service coverage. For instance, Qiao, Leng, et al. [37] introduced a paradigm to jointly manage content placement and delivery in vehicular edge networks. The caching optimization problem was formulated as a double time scale Markov decision process, based on the consideration that content popularity changes are less frequent compared to those associated with vehicle mobility. The authors proposed a deep deterministic policy gradient (DDPG) approach to achieve minimum system costs and content delivery latency. Regarding complex wireless edge networks, research efforts have been devoted to coordinating multiple user devices and improving the efficiency of data services.

Karasik, Simeone and Shamai [38] leveraged the benefits of out-of-band broad-cast D2D communication for caching data delivery in a complex fog radio access network. To minimize the normalized delivery time, a compress-and-forward–based edge caching and D2D communication scheme was proposed, which proved to be an information-theoretically optimal approach. Moreover, researchers have studied the integration of multiple user devices into an edge service node with strong caching capabilities to overcome the weakness of an individual device's constrained storage space. For example, Zhang, Cao, et al. [39] exploited the relations between caching-empowered vehicles in content dispatch services and proposed a social–aware mobile edge caching scheme that leverages deep reinforcement learning to organize social characteristic–aware vehicles in content processing and caching and maximizes dispatch utility.

3.4.3 Hybrid Service–Enabled Edge Caching

Although both infrastructure service-based and D2D service-based edge caching approaches can distribute popular data in proximity to mobile users via local storage capabilities, inherent shortcomings still remain due to the location, caching, and communication characteristics of the edge servers. As mentioned above, the location of the edge cache–enabled infrastructure is fixed, so its data service coverage is difficult to adjust. In D2D edge caching, though user device locations can be flexibly changed, the devices' storage space is usually small. Moreover, due to the large-scale distribution and independent control of different devices, the collaborative caching of multiple devices is sometimes inefficient.

Combining the infrastructure-based approach with the D2D caching approach to build a hybrid edge caching mechanism has emerged as a promising paradigm to address the above-mentioned problems. On the one hand, user devices can spread data to far places as they move, making up for the fixed coverage of infrastructures. On the other hand, an infrastructure uses its large storage capacity to remedy the inability of user devices to store large files. In some cases, cloud data servers are also

Table 3.1 Comparison of edge caching modes

	Infrastructure supported	User device sharing enabled	Hybrid service enabled
Capacity	High	Low	High
Latency	Low	High	Medium
Cost	High	Low	Medium

integrated into data caching systems to provide powerful data source support for the various types of edge servers.

Being a key enabling technique in enhancing data delivery efficiency, the hybrid edge caching mechanism has attracted a great deal of research interest. Wu, Zhang, et al. [40] introduced a D2D-assisted cooperative edge caching scheme in millimeter-dense networks where the cache resources of users and small base stations are jointly utilized for data storage and delivery according to content popularity. Zhao, Liu, et al. [41] designed a caching scheme that combines caching placement and the establishment of D2D with the aid of small base stations. Popular files are prefetched in the local cache during off-peak periods and served to users at peak times, thereby reducing communication pressure on the backhaul link. Zhang, Yu, et al. [42] considered motivations and security in caching service and proposed a blockchain-based cache and delivery market that guarantees the expected reward of both user devices and edge nodes in data sharing. Saputra, Hoang, et al. [43] introduced a proactive cooperative caching approach that uses a content server to predict the content demand for the entire network and that optimizes the distributed caching service of edge nodes. Kwak, Kim, Le and Chong [44] extended the hybrid caching mechanism to the cloud end and proposed a content caching algorithm for joint content caching control in central cloud units and base stations in a hierarchical cellular network.

Table 3.1 compares the key performance of the service modes cited above. Since the infrastructure is deployed and maintained by operators, it has a high capacity and low transmission delay. However, content subscribers need to pay the operators for the caching and transmission services, so infrastructure-supported edge caching usually has a high cost. In contrast, user device sharing–enabled caching utilizes the caching capacity of IoT devices. Although the performance of D2D cache services can be poor, the direct delivery of data without operator participation saves greatly on costs. The hybrid service mode incorporates the previous two approaches and achieves high caching capacity and medium delivery latency, while reducing the cost compared to the infrastructure-supported mode.

3.5 Case Study: Deep Reinforcement Learning–Empowered Social–Aware Edge Caching

To further elaborate the edge caching mechanism and performance evaluation metrics, in this section we present a case study that focuses on deep reinforcement learning–empowered social–aware edge caching management.

3.5.1 System Model

We consider an edge service–empowered vehicular social network with M_i RSUs located in an area i, $i \in \mathcal{I}$. The RSUs' caching space is limited, and their maximum caching capacities are denoted as $\{s_1^r, s_2^r, \ldots, s_{M_i}^r\}$, respectively. In addition, each RSU is equipped with a mobile edge computing server that helps process computation tasks offloaded to it. The computing capacities of these servers are $\{c_1^r, c_2^r, \ldots, c_{M_i}^r\}$, respectively.

The vehicular network consists of smart vehicles. These vehicles generate content, including road congestion status, driving control indications, and vehicle sensing information. This content needs to be processed and delivered among vehicles and roadside infrastructures. The processing and delivery of content are always undertaken jointly and consequently. We use the term *content dispatch process* to represent the combination of content processing and the delivery process. The content involves different data sizes and diverse computing demands. We consider K types of content and denote type k content as $\{f_k, c_k, t_k^{\max}\}$, where f_k and c_k are the content size and required amount of computation, respectively. Variable $t_k^{\max}$ is the maximum latency tolerance of type k content to be dispatched to receiving vehicles. The vehicular network operates within a discrete time model with time slots of fixed length.

To satisfy the computing demand for particular content, the content can either be processed on a vehicle with its on-board computing resources or offloaded to and executed on a mobile edge computing server. Let $x_{v,k}$ be the probability that a given vehicle processes type k content and $1 - x_{v,k}$ the probability that the newly generated content is offloaded to an RSU. Moreover, in the vehicular network, both vehicle-to-vehicle (V2V) and vehicle-to-RSU (V2R) modes can be exploited for content transmission. We use $y_{v,k}$ and $y_{r,m,k}$ to denote the probabilities of a vehicle choosing to deliver type k content through the V2V or V2R mode, respectively, after the content is processed.

In the content dispatch process, social relations and mobility aspects of smart vehicles are jointly exploited. The contact rate is taken as a key indicator to characterize the social relationships between vehicles. Vehicular pairwise inter-contact times follow an exponential distribution. The contact rate between two vehicles in area i is $\lambda_{v,i}$. In addition, RSUs equipped with caching resources can also act as content distribution relays. The various transmission ranges and different locations of these RSUs lead to different contact rates between the RSUs and the vehicles, which are denoted as $\{\lambda_1, \ldots, \lambda_{M_i}\}$, respectively.

3.5.2 Problem Formulation and a DDPG-Based Optimal Content Dispatch Scheme

The content can have different levels of dispatch performance due to various computation offloading targets, caching strategies, and transmission modes. The number of vehicles in an area to which type k content, $k \in \mathcal{K}$, has been dispatched under delay constraint $t_k^{\max}$ can be expressed as

$$n_k^{\text{total}} = x_{v,k} y_{v,k} n_{1,k} + x_{v,k} \sum_m^M y_{r,m,k} n_{2,k,m} + \left(1 - x_{v,k}\right) \sum_m^M \rho_{m,k} n_{3,k,m} \tag{3.5}$$

where $n_{1,k}$, $n_{2,k,m}$, and $n_{3,k,m}$ are the numbers of vehicles that obtain content within their delay constraints through on-board computing and caching, on-board computing with RSU caching, and RSU computing and caching approaches, respectively.

Optimal content dispatching should maximize the total number of vehicles that receive various types of content under specified delay constraints and can be formulated as

$$\max_{\{x_{v,k}, y_{r,m,k}, y_{v,k}, \rho_{m,k}\}} \sum_{k=1}^K n_k^{\text{total}}$$

$$\text{such that} \quad C1: \sum_{k=1}^K x_{v,k} y_{r,m,k} f_k + \sum_{k=1}^K f_k \left(1 - x_{v,k}\right) \cdot \rho_{m,k} \le s_m^r, \quad m \in M$$

$$C2: \ 0 \le x_{v,k}, y_{v,k}, y_{r,m,k}, \rho_{m,k} \le 1, k \in \mathcal{K}, m \in M \tag{3.6}$$

$$C3: \sum_{m=1}^M y_{r,m,k} + y_{v,k} = 1, \quad k \in \mathcal{K}$$

$$C4: \sum_{m=1}^M \rho_{m,k} = 1, \quad k \in \mathcal{K}$$

where constraint C1 indicates that the amount of data cached on RSU m should not exceed its cache capacity; C2 gives the ranges of the decision variables $x_{v,k}$, $y_{v,k}$, $y_{r,m,k}$, and $\rho_{m,k}$; C3 implies that either content should be transmitted V2V or an RSU should be selected for V2R data delivery; and constraint C4 indicates that one of the M RSUs should be selected when edge computing services are utilized to process content.

To address this problem, we design a DDPG-based dispatch scheme by leveraging a deep reinforcement learning approach. The DDPG is a policy gradient deep reinforcement learning algorithm that concurrently learns policy and value functions in the learning process. The DDPG agent learns directly from unprocessed observation spaces through a policy gradient method that estimates the policy weights,

Fig. 3.3 Architecture of the
DDPG-based scheme

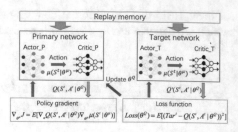

and employs an actor–critic model to learn the value function and update the actor model. Since the DDPG utilizes a stochastic behavior policy for strategy exploration but estimates a deterministic target policy, it greatly reduces learning complexity.

In our designed DDPG learning approach, the action set taken in slot t is $\mathcal{A}^t = \{x_{v,k}^t, y_{r,m,k}^t, y_{v,k}^t, \rho_{m,k}^t, \alpha_k^{t,i}, p_{i,j,k}^t\}$, where $k \in \mathcal{K}$, $m \in \mathcal{M}_i$, and $i \in \mathcal{I}$. The state at time slot t is $\mathcal{S}^t = \{S_{1,k,i}^t, S_{2,k,m,i}^t, S_{3,k,m,i}^t, S_{4,k,i}^t, S_{5,k,m,i}^{t,j,i}, S_{6,k,m,i}^{t,j,i}\}$, which represents the number of vehicles that have obtained content through various ways. Moreover, we introduce two neural network parameters, namely, θ^μ and θ^Q, in the DDPG learning process. The parameter θ^μ is updated by the primary actor neural network using the sampled policy gradient, which can be shown to be

$$\nabla_{\theta^\mu} J = \mathrm{E}[\nabla_a Q(S^t, A^t|\theta^Q) \nabla_{\theta^\mu} \mu(S^t|\theta^\mu)] \tag{3.7}$$

where $Q(S^t, A^t|\theta^Q) = \mathrm{E}[u^t + \eta Q(S^{t+1}, \mu(S^{t+1})|\theta^Q)]$ is an action value function, and $\mu(S^t|\theta^\mu)$ is the explored policy. The term θ^Q is updated by a primary critic neural network by minimizing a loss function, which is defined as

$$Loss(\theta^Q) = \mathrm{E}[(Tar^t - Q(S^t, A^t|\theta^Q))^2] \tag{3.8}$$

where the target value $Tar^t = u(S^t, A^t) + \eta Q'(S^{t+1}, \mu'(S^{t+1}|\theta^{\mu'})|\theta^Q)$. The DDPG-based content dispatch scheme concurrently learns an action value function and dispatch policies. It uses off-policy data and the Bellman equation to learn the action value function, and it utilizes the action value function to learn the policy.

Figure 3.3 shows the architecture of the DDPG-based dispatch scheme. The compositions of the primary and target networks are similar, and both have an actor and a critic. The actor and critic are two deep neural networks. In the primary network, the actor explores the content dispatch policy $\mu(S^t|\theta^\mu)$, while the critic helps the actor learn a better policy through a gradient approach. The target network is an old version of the primary network. It generates a target value to train the critic in the primary network, where the policy θ^Q is updated through the calculated function $Loss(\theta^Q)$. The replay memory stores the learning experience used to update the actor and critic parameters.

3.5.3 Numerical Results

We evaluate the performance of the proposed content dispatch schemes based on a real traffic dataset that contains mobility traces of approximately 500 taxi cabs in the San Francisco Bay Area.

Figure 3.4 shows the convergence of the proposed DDPG-based dispatch scheme with different values of λ_v, the average vehicular contact rates of the different combinations of areas. In different scenes with different λ_v values, the DDPG-based scheme converges within 8,000 iterations. In addition, the figure indicates that λ_v significantly affects the number of vehicles that can receive content within the specified constraint. A larger λ_v means a higher probability of vehicles meeting and interacting with each other. Content dispatch efficiency improves as λ_v increases. It is worth noting that, in practice, the proposed DDPG-based dispatch scheme is executed offline. For a given steady vehicular traffic flow, we could obtain the optimal content dispatch strategies for vehicles with various states in advance. The strategy set is stored in the control center and can be accessed and applied to vehicles directly, without very many learning iterations.

Figure 3.5 compares the content dispatch performance of the proposed DDPG-based scheme under two social–aware data forwarding approaches, that is, Epidemic and ProPhet. In the Epidemic scheme, a vehicular data carrier forwards data to its contacted vehicle in V2V mode if the target vehicle does not have the data. In the

Fig. 3.4 Convergence of the proposed DDPG-based dispatch scheme

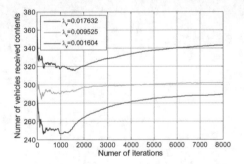

Fig. 3.5 Comparison of the content dispatch performance under different approaches

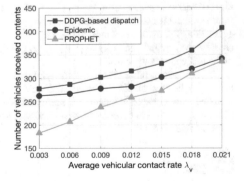

ProPhet scheme, the vehicular carrier only forwards the data if its contacted vehicles have higher contact rates compared to its own. We use the number of vehicles that can receive content under their delay constraints as the metric to evaluate the caching performance. It is worth nothing that this metric can be considered a special form of the hit rate ratio metric. The figure shows that the DDPG scheme is the most efficient, since it jointly exploits the data dispatch capabilities of both vehicles and RSUs, while adaptively optimizing dispatch strategies according to the various vehicular contact rates of different areas. In contrast, under the Epidemic approach, vehicles forward cached content to contacted vehicles only if their caching space is unoccupied. This approach uses only V2V delivery and ignores the data distribution capabilities of RSUs. In ProPhet, a vehicle only forwards cached content if the contacted node has a contact rate higher than its own. Consequently, ProPhet relies only on RSUs for content distribution when λ_v is small, but relies on V2V content delivery when λ_v exceeds the contact rate metrics of the RSUs. The single data delivery mode of each of these two approaches seriously decreases their dispatch efficiency.

Chapter 4
Mobile Edge Computing for Beyond 5G/6G

Abstract This chapter first introduces the mobile edge computing (MEC) paradigm in beyond 5G and 6G networks. The motivations, applications, and challenges of integrating MEC into 6G are discussed in detail. We then present a new paradigm, MEC-empowered edge model sharing, as a use case for 6G. Furthermore, the architecture and processes of the MEC-empowered edge model sharing system are provided to show the integration angles of MEC and 6G networks.

4.1 Fundamental Characteristics of 6G

Although still in the early stage, a number of studies have provided visions for 6G [45–47]. Besides considerably improved data rates and communication latency, 6G networks are also considered to be human-centric and connected intelligence. The key features of 6G networks should be as follows.

- *Extremely high data rates and low latency*: Applications in 6G require much higher data rates and much lower latency than in 5G. The data transmission rates are expected to be in the hundreds of gigabytes or even terabytes. The latency should be extremely reduced, and services and applications are thus provided in real time. The extremely high data rates also generate new requirements for more spectrum resources. Hybrid terahertz–visible light communication systems are expected to offer more unexplored bandwidth resources for 6G networks.
- *Low energy consumption*: The increasing number of connected smart devices, such as Internet of Things devices and smartphones, in 6G requires the energy consumption to be low to extend their running time and provide reliable services.
- *High edge intelligence*: Artificial intelligence (AI) is assumed to play a crucial role in 6G networks. The concept of connected devices has evolved into connected intelligence in 6G. Edge intelligence is a key enabler of 6G networks [48]. Network performance will be improved by optimizing the allocation of resources such as the spectrum, computation, and power in the network [49]. Moreover, the integration of AI techniques into edge networks is expected to be greatly improve the quality of service (QoS).

© The Author(s) 2022   37
Y. Zhang, *Mobile Edge Computing*, Simula SpringerBriefs on Computing 9,
https://doi.org/10.1007/978-3-030-83944-4_4

- *High security and privacy*: As the number of involved users increases dramatically in 6G, their devices generate a large amount of data. Since the generated data contain users' private information, the risk of data leakage during data transmission and storage is a major threat for 6G networks. Emerging technologies such as blockchain and federated learning are needed to enhance network security and data privacy.

4.2 Integrating Mobile Edge Computing (MEC) into 6G: Motivations, Applications, and Challenges

In cloud-based scenarios, the long-distance transmission of data from end devices or edge servers to the cloud incurs great latency and security risks and consumes a great amount of bandwidth. In 6G systems, a series of emerging applications, such as virtual reality (VR) and real-time video, require ultra low latency performance. Meanwhile, the explosive growth of smart devices in 6G also brings a large amount of distributed computational resources to the edge. In this regard, conventional cloud-based computation can hardly satisfy the expected performance requirements of 6G systems.

4.2.1 Use Cases of Integrating MEC into 6G

MEC enables the computation of applications and services to be executed at the edge of networks, reducing transmission latency and mitigating the threat of data leakage. Moreover, by deploying AI algorithms on edge servers, MEC leverages the distributed computational capabilities of devices and enables edge intelligence to be extensively realized in 6G. Thus, MEC is a key enabling technology for 6G systems. In areas that benefit from MEC, the use cases of MEC can be classified into three categories: consumer-oriented services, operator and third-party services, and network performance and quality of experience improvement services [50, 51].

In the category of consumer-oriented services, end users benefit from MEC by offloading computation to an edge server to run various 6G applications that require high computational capability and low latency performance. For example, in the scenarios illustrated in Fig. 4.1, such as face recognition or smart camera applications, the end devices need to analyze collected images in near real time. In such a case, neither cloud servers nor end devices can satisfy the requirements, due to long transmission distances or constrained computation resources. MEC enables end devices to run such low latency applications by offloading heavy computation to edge servers.

In the use cases of operators and third-party services, operators and third parties benefit from MEC systems. In 6G networks, the increasing numbers of smart devices generate huge amounts of distributed data. Directly transmitting these data

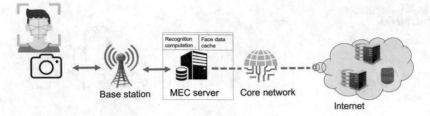

Fig. 4.1 Example application of MEC: Face recognition

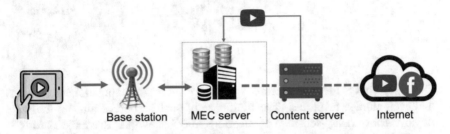

Fig. 4.2 Example application of MEC: video content caching

to the cloud will occupy a great deal of communication resources and lead to the additional consumption of storage and computation resources on cloud servers. In such scenarios, the MEC server operates as the gateway to collect and process generated data in advance. The processed data are then transmitted to cloud servers by the MEC server, which significantly reduces the transmission load from the edge to the cloud and mitigates the computational burden of centralized cloud servers.

In terms of network performance and quality of experience improvement services, MEC alleviates the congested backhaul network by means of content caching and traffic optimization. In content caching, MEC servers store popular content in advance by analyzing the historical records of users in their area, as shown in Fig. 4.2. Once the users request related content, the MEC servers will return the cached content directly to them. Through the content caching of MEC applications, the transmission latency is reduced and the user experience is improved. Moreover, MEC can help to optimally schedule traffic by gathering and analyzing network information and user requirements at the edge.

4.2.2 Applications of Integrating MEC into 6G

Considering the above benefits, MEC can be applied in a series of 6G applications. As shown in Fig. 4.3, the applications can be categorized as follows.

- *Dynamic resource sharing*: In 6G networks, the increasing numbers of connected devices and delay-sensitive applications require tremendous resources to ensure

Fig. 4.3 Applications of
MEC for 6G

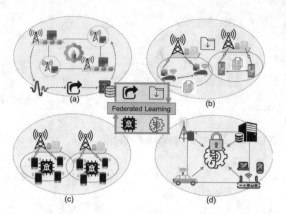

QoS. The types of resources include spectrum resources, computational resources, and even storage resources. The limitation of such resources hinders the wide deployment of delay-sensitive applications in 6G networks. Resource sharing is an effective way to mitigate resource constraints. However, what to share and how to share are two basic issues that must be carefully addressed in resource sharing. MEC provides solutions to these issues by modeling and analyzing the network and optimizing sharing policies.

- *Distributed device-to-device caching*: In 6G networks, massive amounts of high-quality low latency applications, such as online games and real-time multimedia, generate huge amounts of content on edge devices. Instead of storing these contents on the cloud server (e.g., a macro base station), caching these contents at the edge considerably reduces the transmission costs and centralized storage burden. In 6G, since the computational and storage capabilities of smart devices will be significantly improved, caching content with end users can better leverage the distributed resources to reduce transmission latency and improve the QoS. End users with constrained resources are caching requesters, while end users with sufficient resources are caching providers. The device-to-device caching system is illustrated in Fig. 4.3b. The MEC server collects the information of end users under its coverage and optimally determines the caching strategy by analyzing and predicting content popularity among distributed users. The analysis can be conducted through optimization algorithms and AI algorithms that jointly consider the latency requirements and current information on the demands and offers from end users.

- *Joint edge computation offloading*: Since blockchain maintenance and the aggregation of updates require intensive computation, it is a challenging task for edge servers to execute computations within the applicable constraints, especially with large numbers of participating nodes. To alleviate the computational pressure, MEC completely utilizes the distributed computing resources by splitting the computation task into shards and offloading these onto other computing servers with sufficient computing resources. Moreover, the offloaded computation tasks can also leverage the target user's data to complete the computation, which can further reduce the transmission overhead. An overview of the computation offloading scheme is shown in Fig. 4.3c.

- *Secure and private data analysis*: In 6G networks, a large amount of network data must be processed and analyzed to improve the QoS. With increasing concerns of data security and privacy, conventional cloud-based mechanisms raise serious threats of the leakage of user data. MEC allows the data to be analyzed at the edge of networks or even at the side of end users. Empowered by emerging paradigms such as federated learning [52], MEC will considerably enhance data privacy in the data analysis of 6G applications.

4.2.3 Challenges of Integrating MEC into 6G

Although integrating MEC into 6G has a series of benefits, new challenges also arise. Considering the characteristics of 6G networks and connected devices, the main challenges can be summed in three points: the heterogeneity of distributed resources, the high mobility of end users such as vehicles and mobile devices, and increasing security and privacy concerns.

- *Distributed heterogeneous resource management*: In 6G networks, a huge amount of multidimensional heterogeneous resources have emerged as the number of smart devices has increased. In addition, as the capabilities of mobile devices improve, many resources are distributed among these devices. To improve the QoS and utility of distributed resources, heterogeneous resources need to be optimally allocated in real time. MEC plays a crucial role in edge resource management. However, the heterogeneity of the distributed resources, the dynamic system states, and critical latency constraints raise new challenges to integrating MEC into 6G for real-time resource management. Ways to improve the intelligence of an MEC system to address resource heterogeneity and to improve latency performance for real-time resource allocation require further investigation.
- *Reliability and mobility*: There are many fast-moving scenarios in 6G network, such as vehicular networks and mobile networks. In these scenarios, end users are continuously moving in the network. The network topology therefore varies, since the times and communication channels between users and base stations are unstable. However, the demand for low latency and high-reliability services also exists among end users. In a conventional MEC system, the MEC server executes computation tasks or caches content to reduce the transmission delay and improve computational capability. New MEC schemes must therefore be developed for 6G to guarantee the continuity of services for moving users in dynamic networks.
- *Security and privacy*: The increase in the number of end devices also generates huge amounts of user data. Leveraging these data for analysis can improve the QoS. For example, the accuracy of advertising recommendations can be further improved by learning the behaviors of users. Moreover, using AI algorithms to

learn the network running data can help to improve the network performance to satisfy the requirements of 6G networks. However, these data can contain sensitive user information. The risks of data leakage increase in this process. To integrate MEC into 6G, concerns of user privacy and security need to be addressed. More privacy-preserving machine learning algorithms and security collaboration mechanisms are required to enhance the security and privacy of MEC systems.

4.3 Case Study: MEC-Empowered Edge Model Sharing for 6G

4.3.1 Sharing at the Edge: From Data to Model

In conventional data sharing scenarios, the data providers share original data directly with the data requesters, which incurs a large amount of data transmission and increases the risk of data leakage. For example, a conventional traffic prediction application scenario is depicted in Fig. 4.4a. Distributed cameras share their video data with others and the cloud server to obtain overall traffic flow conditions. The traffic analysis and prediction are executed on the cloud server and then sent back to the end users. In the model sharing scenario, shown in Fig. 4.4b, end users equipped with MEC servers train locally based machine learning models with their collected video data. The trained machine learning models are shared with other users requesting the sharing of traffic data. Requesters then run the received machine learning model on their local data and build a new model for predicting real-time traffic conditions. By leveraging MEC to share the computing model instead of original data at the edge, response latency is reduced and data privacy is considerably enhanced.

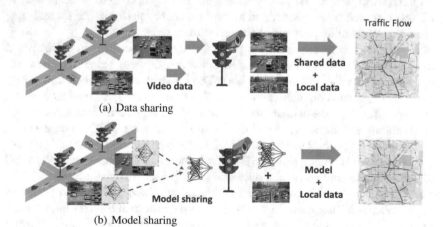

Fig. 4.4 From data sharing to model sharing

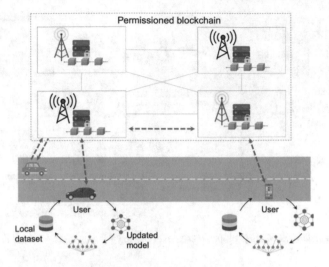

Fig. 4.5 The architecture of MEC-empowered model sharing

4.3.2 Architecture of Edge Model Sharing

The architecture of edge model sharing is illustrated in Fig. 4.5. We introduce blockchain into the proposed architecture to construct a secure sharing mechanism among end users who lack mutual trust. In the proposed sharing scheme, the providers register to a permissioned blockchain with their data profiles and run local training on their data to build the machine learning models. The permissioned blockchain runs on the base stations or roadside units as the parameter server. The registration information of users, the model parameters, and the sharing events are recorded in the blockchain. The requesters retrieve the blockchain for potential providers and request multiple users to provide the models. Through blockchain, the providers can be rewarded for sharing their models with requesters.

4.3.3 Processes of Edge Model Sharing

Based on the proposed architecture, edge sharing applications can be performed in MEC systems. The overall edge sharing procedures are shown in Fig. 4.6, which shows all the processes of MEC-empowered model sharing. The detailed processes are as follows.

1. *Initialization*: When data providers join the system, local similarity clustering is performed to classify these datasets, as well as the providers, into various categories. The similarity between different datasets is quantified by their logical distances, such as cosine similarity and Jaccard similarity. For a specific data

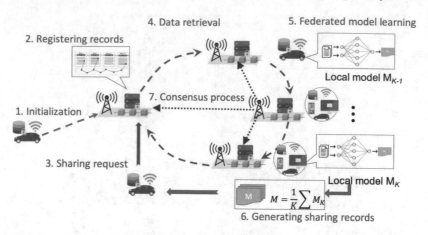

Fig. 4.6 The processes of MEC-empowered model sharing

provider P_i, nearby blockchain nodes will search the blockchain network to find
similar data records. Then the ID of the dataset from P_i is generated based on
the hash vectors of similar records, to ensure that similar datasets hold close
IDs. The participants are divided into different communities according to their
ID distances, that is, data similarity.

2. *Registering retrieval records*: Data provider P_i is required to register in the
   blockchain by sending a public key PK_r and its data profiles to a nearby
   blockchain node (e.g., MEC server). The blockchain node then generates a data
   retrieval record for provider P_i and broadcasts it to other nodes in the network
   for verification. The nodes in the blockchain verify their received records and
   pack them into candidate blocks. The candidate blocks are then verified through
   a consensus protocol and written to the permissioned blockchain if they are
   verified.

3. *Launching data sharing requests*: Data requester P_r submits a sharing request
   $Req = \{f_1, f_2, ..., f_x\}$ that contains the requester ID, the requested data category, and the time stamp to a nearby blockchain node SN_{req}. The sharing request
   Req is signed by P_r through the requester's private key SK_r.

4. *Data retrieval*: When the blockchain node near P_r receives the sharing request,
   it first validates that the identity of P_r is legal. If P_r is an authorized user, the
   blockchain node searches for the sharing records in the permissioned blockchain
   to check whether the request has been processed before. If there is a hit, the cached
   model will be returned to requester P_r directly. Otherwise, the blockchain node
   will carry out a data retrieval process among registered providers according to
   their ID distances, to find related data providers.

5. *Data model learning*: Data providers related to the request Req work together to
   train a collaborative data model $\mathcal{M}$. The local training samples consist of a request
   query f_x and its corresponding query results $fx(D)$, $D^T =< f_x, f_x(D) >$. The
   local models are trained on dataset D^T and aggregated into a global model $\mathcal{M}$.

The learned global model is then returned to the requester P_r as the result, which is also cached by the system for future requests. The requester can obtain the exact results it required based on the received model and its local data.

6. *Generating sharing records*: Data sharing events are recorded in the blockchain as transactions and broadcast to other blockchain nodes for verification. These records are collected by blockchain nodes and packed into candidate blocks.
7. *Carrying out consensus*: Candidate blocks consisting of data sharing records are verified by blockchain nodes participating as data providers. The blockchain nodes compete for the opportunity to generate blocks of the blockchain though consensus protocols such as proof of work or delegated proof of stakes. Nodes that obtain the right to generate blocks add their candidate blocks to the blockchain. The sharing records in the blockchain are traceable and tamper proof.

The combination of blockchain and federated learning enables secure intelligent data sharing in 6G networks. Based on federated learning, the data sharing among mobile devices is transferred to model sharing, which avoids the transmission of original data and reduces the risks of data leakage. Moreover, integrating the training process in federated learning with the blockchain consensus process improves the utilization of computing resources and the efficiency of data sharing. This edge model sharing case shows the great potential of integrating MEC into 6G networks to improve QoS and applications. MEC brings edge intelligence to wireless edge networks and enhances the connected intelligence among end devices in 6G networks.

Chapter 5
Mobile Edge Computing for the Internet of Vehicles

Abstract The advancement of cyber physical information has led to the pervasive use of smart vehicles while enabling various types of powerful mobile applications, which usually require high-intensity processing under strict delay constraints. Given their limited on-board computing capabilities, smart vehicles can offload these processing tasks to edge servers for execution. However, a highly dynamic topology, a complex vehicular communication environment, and edge node heterogeneity pose significant challenges in vehicular edge computing management. To address these challenges, in this chapter we investigate the characteristics of edge computing from both the application and service perspectives and introduce a hierarchical edge computing framework. Moreover, we leverage artificial intelligence technology to propose efficient task offloading and resource scheduling schemes.

5.1 Introduction

Due to the promising advancements of Internet of Things technology and wireless communications, smart vehicles empowered with environmental perception, information processing, and automatic control capabilities have emerged. Smart vehicles bring us powerful vehicular applications, such as autonomous driving, voice recognition, and car entertainment, and help to build a smarter, safer, and more sustainable transportation system. These applications usually require intensive computational processing. However, constrained by on-board computing resources, an individual smart vehicle might not provide sufficient computing power, which makes it difficult to ensure that application tasks are completed on time.

Mobile edge computing (MEC) provides a feasible approach to the above problem. By deploying computing servers in vehicular access networks, application tasks can be offloaded to the network edge for efficient execution. The offloading process leverages the wireless links between smart vehicles and roadside units (RSUs) for task data delivery and the acquisition of processing results. Moreover, smart vehicles that have spare computing resources can be exploited as edge computing servers to serve adjacent vehicle task generators in vehicle-to-vehicle (V2V) communication [53]. To specifically describe this edge computing approach that uses vehicular communication in task offloading, we call it vehicular edge computing (VEC).

© The Author(s) 2022
Y. Zhang, *Mobile Edge Computing*, Simula SpringerBriefs on Computing 9,
https://doi.org/10.1007/978-3-030-83944-4_5

In a VEC system, the high-speed movements of vehicles and rapid changes in network topology lead to unique characteristics that, unlike traditional edge computing systems, are designed for handheld mobile smart terminals. Moreover, these characteristics lead to new challenges and require the implementation of key techniques in the MEC architecture design, computing service scheduling, and resource management, which are investigated and described as follows.

5.2   Challenges in VEC

Identifying the technical challenges of VEC design and management is a prerequisite for optimal edge computing services. According to the characteristics of road traffic environments and vehicular edge networks, we summarize the challenges into four items.

- *A highly dynamic network topology and unstable service relationships*: The dynamic changes of the network topology due to high-speed vehicle movement is the most important feature of VEC. This topology change can greatly affect transmission rates, interference, energy consumption, and so on. Since communication plays a key role in VEC task offloading, a dynamic topology implies complicated wireless access point switching, power adjustments, and interference suppression for edge service management. Moreover, considering the limited coverage of densely deployed base stations (BSs) in 5G/beyond 5G networks, high-speed moving vehicles can leave the communication range of a BS within a short time. When a high-speed vehicle generates a task with an intensive computing demand, it is difficult for a single VEC server equipped on a BS to complete the calculation process within the time the vehicle remains within the BS's coverage. Unstable service relations are thus induced between VEC servers and users, which further complicates the VEC management mechanism.
- *Strict low latency constraints and large amounts of task data*: Most of Internet of Vehicles applications are related to autonomous driving control and traffic safety improvement, which always have strict low latency constraints. For example, a vehicle's reaction time to a suddenly appearing obstacle needs to be limited to milliseconds. Thus, the fast and efficient processing of obstacle identification and of control instruction generation becomes a necessary prerequisite. This requires edge servers to provide sufficient computing resources. However, on congested roads with a large number of vehicles, adequate serving capacities are often difficult to achieve. Furthermore, as mentioned before, edge computing services rely on task data transmission between user nodes and servers. In autonomous driving applications, vehicular sensors, such as cameras, millimeter wave radar, and lidar, continue to generate large amounts of data, which seriously challenges the communication capabilities of vehicular networks.
- *Heterogeneous and complex communications*: Vehicular networks consist of smart vehicles, RSUs, and BSs. These devices and infrastructures form a variety of communication relationships, including V2V, vehicle-to-RSU (V2R), and vehicle-to-

infrastructure (V2I), which are collectively referred to as vehicle-to-everything, or V2X. Diverse V2X communications can work in different frequency bands or share the same spectrum resources. In addition, consistent sets of standards have been created for the deployment and operation of vehicular communication, such as Dedicated Short Range Communication in the United States, Cooperative Intelligent Transport Systems in Europe, and IEEE 802.11bd and 5G New Radio V2X in the 5G era. Large-scale heterogeneous devices following multiple types of standards communicate in parallel in constrained frequency bands, which makes vehicular communication extremely complicated and leads the efficient task offloading a challenge.

- *Decentralized and independently controlled edge service nodes*: Empowered with a processor, cache, and communication interface, a smart vehicle can be considered a mobile edge server when it has surplus computing resources and helps other vehicles through V2V task offloading. For application tasks with highly intensive computing requirements, the capability of a single vehicular server might not meet demands. In such a case, aggregating multiple vehicular servers to form a group entity with powerful service capabilities is a promising approach to the problem. However, since the vehicles in the network are mobile and distributed, a centralized control mechanism is spectrum inefficient and time-consuming. Furthermore, each vehicle's service willingness and driving behavior are independently controlled by its owner. It is impractical to request that all vehicle owners follow scheduling instructions unconditionally.

5.3 Architecture of VEC

To address the challenges mentioned, we propose an architecture of VEC to guide VEC service management that illustrates the system components and their logical relations.

Figure 5.1 shows the proposed architecture, which is divided into four layers. The bottom layer is the application layer. It consists of the smart vehicles and powerful vehicular applications. These vehicles have varied computing and communication capabilities while driving throughout large areas with different speeds and route plans. The applications they run, such as autonomous driving and navigation, can differ in terms of computing resource requirements and delay constraints. Vehicle characteristics and application requirements can be used as the input of the upper layers, which drive the edge service strategy adjustment.

In the edge server layer, there are three types of vehicular serving infrastructures. The first one consists of BSs equipped with MEC servers. The BSs can be macro BSs, micro BSs, or even pico BSs. They use V2I transmission to gather the computation tasks of the vehicles traveling in the coverage area, send the tasks to the MEC server for processing, and finally return the results to the vehicles. Similar to the BS operation, RSUs equipped with MEC servers are deployed along roads, serving the vehicles traveling past them.

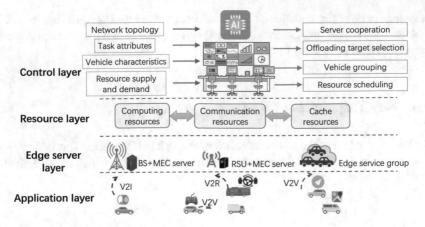

Fig. 5.1 Architecture of a VEC system

It is worth discussing the edge service group formed by multiple smart vehicles with idle computing resources. The group members can be either stationary vehicles in a parking lot or moving vehicles on roads. The offloaded computing task needs to be shared by multiple vehicles, and the execution of each part of the task usually requires the close cooperation of the other parts. The communication capabilities among team members should be efficient and reliable. A service group is thus usually formed by vehicles geographically adjacent to each other.

At the resource layer, the edge serving resources provided by BSs, RSUs, MEC servers, and smart vehicles are logically divided into three categories: computing resources, communication resources, and cache resources. All the types of resources work cooperatively in task offloading and processing. The communication resource is responsible for task transmission and the delivery of calculation results, while the cache resource helps store task data in the servers. It is worth noting that, in some cases, cross-type collaboration can be implemented between heterogeneous resources. Tasks generated in an area with poor computing service but sufficient bandwidth can be transmitted to a remote area with powerful computing capabilities. This case can be viewed as using the cost of communication resources in exchange for computing resources.

The control layer is at the top of the architecture, monitoring the service states of the edge system and deriving optimal scheduling strategies. More specifically, the control units gather data on network topology, task attributes, vehicle charac-teristics, and resource states. The gathered information is then input into an AI module to analyze service supply and demand trends. Based on the analysis results, the units are used to form an effective management plan that determines offloading target servers, decides the multi-vehicle grouping mode, coordinates the interaction between heterogeneous servers, and optimizes various types of edge resources. From the implementation perspective, the control module can be a centralized control entity in charge of an entire network or distributed controllers equipped on BSs or RSUs that are responsible for scheduling service resources within a local area, or even a head vehicle in vehicular groups.

5.4 Key Techniques of VEC

Under the guidance of the architecture, many works have focused on several key technical issues in VEC construction, management, and operation, which are investigated in the following.

5.4.1 Task Offloading

Task offloading is the essential motivation for the proposed edge service, and it is also the core function of the VEC system. Since offloading processes can have diversified optimization goals under different application scenarios, there are a variety of corresponding offloading mechanisms.

To reduce energy bills and create green edge systems, energy efficiency has been considered an optimization goal in many studies. The energy consumption of task offloading is mainly split into two parts: consumption in data communication and task processing. Lower radio power and shorter communication times can reduce transmission energy costs. Based on a signal fading model and Shannon's theorem, offloading target servers with smaller transmission distances and less wireless interference must be chosen. Regarding the processing energy part, different devices have diverse energy efficiency features. For example, given the use of many different types of silicon chips, the energy cost of a unit calculation performed by an on-board processor is usually higher than that of the dedicated processor in an MEC server. Thus, without significantly increasing the communication energy overhead, offloading tasks from a vehicle to an edge server usually improves the system's overall energy efficiency.

Many works have been devoted to optimizing offloading energy efficiency. Pu, Chen, et al. [54] introduced a hybrid computing framework that integrates an edge cloud radio access network to augment vehicle resources for large-scale vehicular applications. Based on this framework, a Lyapunov-theoretic task scheduling algorithm was proposed to minimize system energy consumption. Zhou, Feng, Chang and Shen [55] leveraged vehicular offloading to alleviate the energy constraints of in-vehicle user equipment with energy-hungry workloads. They designed a consensus-based alternating direction driven approach to determine the optimal portion of tasks to be offloaded to edge servers. Li, Dand, et al. [56] modeled task offloading process at the minimum assignable wireless resource block level and presented a measure of the cost-effectiveness of allocated resources to help reduce the required offloading energy.

Another optimization goal is the quality of experience (QoE), which has drawn great interest recent years. The QoE reflects users' satisfaction with the task offloading performance, and it can be quantified in several ways. One way is through the offloading delay. In most cases, users want tasks to be completed as quickly as possible. To meet this demand, both the task data transmission delay and computation delay should be minimized. However, because of vulnerable communications between user vehicles and RSUs, as well as task congestion at edge servers, guaranteeing timely offloading is a challenge. The division and distributed execution of computing tasks, which reduce the transmission channel requirements and server loads, has become a promising paradigm for addressing this challenge. Ren, Yu, He and Li [57] leveraged the collaboration of cloud and edge computing to partially process vehicular application tasks. They further proposed a joint communication–computation resource allocation scheme that minimizes the weighted-sum task offloading latency. Lin, Han, et al. [58] took a software-defined networking approach to edge service organization and introduced a distributed delay-sensitive task offloading mechanism over multiple edge server–empowered BSs.

Edge service reliability is also a key concern of QoE. However, the dynamic and uncertain vehicular environments create critical challenges in preserving user satisfaction. Many works have addressed this challenge. Ku, Chiang and Dey [59] focused on providing sustainable computing services for edge infrastructures powered by solar energy. Through offline solar energy scheduling and online user association management, the risk of power deficiency and intermittent edge service have been reduced. To ensure the high reliability of completion of vehicular application tasks, Hou, Ren, et al. [60] utilized both partial computation offloading and reliable task allocation in a VEC system, and proposed a fault-tolerant particle swarm optimization algorithm for maximizing computing reliability under delay constraints. To minimize long-term computing quality loss in unpredictable network states, Sun, Zhao, Ma and Li [61] formulated a nonlinear stochastic optimization problem to jointly optimize radio allocation and computing resource scheduling.

In addition to service reliability, cost is an important factor in user QoE. From the different perspectives of VEC operators and users, cost has distinct measurement approaches. Operators mainly consider the cost of the deployment of service facilities. To fully cover an area, a large number of MEC servers could be required, significantly increasing edge construction costs. To address this problem, Zhao, Yang, et al. [62] used unmanned aerial vehicles to act as relay nodes in forwarding computation tasks between smart vehicles and MEC servers. In this way, the service coverage of a single MEC server is improved, and both the number and construction cost of servers are reduced. Users, on the other hand, are concerned about minimizing the costs of using edge services. For instance, Du, Yu, et al. [63] made full use of TV white space bands to supplement the bandwidth for task offloading and introduced a cognitive vehicular edge networking mechanism that minimizes the communication costs of vehicular terminals. Deng, Cai and Liang [64] leveraged multi-hop vehicular ad hoc networks in task offloading and found the multi-hop routing path with the lowest costs through a binary search approach.

With the severe increase in malicious attacks and eavesdropping, security is an increasingly important issue. In VEC systems, due to the various ownership and management strategies of different vehicles and edge servers, it is difficult to ensure that all participating edge service entities are trustworthy. Consequently, security protection mechanisms and privacy preservation measures need to be implemented. Hui, Su, Luan and Li [65] focused on securing vehicle cooperation in relaying service requests and designed a trusted relay selection scheme that identifies relay vehicles by their reputations. Blockchain technology, a tamper-resistant distributed ledger of blocks that keeps data in a secure manner, is attracting growing attention and has been adopted in VEC. For instance, Liu, Zhang, et al. [66] presented a blockchain-empowered group authentication scheme that identifies vehicles distributively based on secret sharing and a dynamic proxy. To protect data privacy in task offloading, Zhang, Zhong, et al. [67] proposed a fuzzy logic mathematical authentication scheme to select edge vehicles, maintaining sensitive communications and verifications only between vehicles.

5.4.2 Heterogeneous Edge Server Cooperation

Although VEC is a promising paradigm for alleviating the heavy computation burden of smart vehicles, an individual edge server is still resource constrained, raising several challenges in the pervasive and efficient deployment of edge services. On the one hand, limited computing power and energy supply make it hard for servers to complete complex tasks under delay constraints. On the other hand, the wireless communication range of the BS or RSU on which the server depends is limited, which further constrains edge service capabilities from the communication perspective. An effective way to resolve these problems is the use of multi-server collaboration. Considering the different types of edge servers in VEC, combining these heterogeneous servers into a joint service leads to a variety of collaboration modes and approaches, which are illustrated as follows.

Cooperation among multiple edge servers equipped on communication infrastructures is the most widely used mode [68]. This mode benefits greatly from the wire connections between infrastructures, such as optical fiber and large capacity twisted pair cable, through which the task data can be transferred and exchanged between multiple servers at high speed and low cost [69]. However, the different computing capabilities of these servers pose challenges in selecting the offloading server and computing resource allocation. Whether to split a large task into multiple subtasks and distribute them in several parallel servers or to merge multiple small tasks into a few large tasks to run on selected servers needs to be carefully designed [70]. In the parallel execution mode, the inefficiency of the entire task processing due to the lag of a server of poor capability is also a key issue to be addressed. In addition, we need to optimize how tasks are offloaded from vehicles to servers. For example, task data can be collected through a BS and then spread to other BSs and servers using the wired connection between BSs. One could also use concurrent wireless delivery between multiple vehicles and the RSUs they can access.

Groups of multiple smart vehicles providing sufficient service capabilities to other user vehicles are another mode of edge service collaboration [71]. This mode makes full use of unoccupied on-board computing resources and is characterized by flexible organization and pervasive availability. However, challenges still exist in the efficient implementation of inter-vehicle service collaboration. The most serious challenge comes from the independence of the different vehicles. The vehicles are owned and controlled by different persons, with various driving route plans and degrees of service willingness. In addition, these vehicles can differ in terms of their idle resource capacity, maximum communication distance, and server energy supply [72]. This independence brings complexity and uncertainty to the vehicular server collaboration. Moreover, the number and geographic distribution of cooperative vehicle servers are also key factors that should be taken into account, since the vehicle distribution density will affect the trade-off between the computing service capacity and spectrum multiplexing efficiency. Thus, vehicular service collaboration schemes with efficient server grouping, resource scheduling, and vehicle owner incentives are required.

Integrating the servers equipped on infrastructures with on-board servers produces a heterogeneous edge service collaboration mode [73]. This mode takes full advantage of the large coverage and strong capabilities of infrastructure servers and leverages on-board servers to make up for the lack of flexibility of the infrastructures. In this mode, taking into account the advantages of V2V communication in terms of small path loss and low transmission delay, tasks with light loads and strict delay constraints are offloaded to on-board servers, while tasks of high computational intensity and loose delay constraints are usually offloaded to infrastructure servers. In case the edge servers cannot meet the vehicular task demands, the number and scope of collaborative servers can be expanded; that is, three-level coordination consisting of cloud servers, infrastructure servers, and on-board servers can be jointly scheduled in matching various types of application tasks.

5.4.3   AI-Empowered VEC

In recent years, we have witnessed unprecedented advancements and interest in artificial intelligence (AI). Machine learning, a key AI technology, provides entities working in complex systems the ability to automatically learn and improve from experience without being previously and explicitly programmed.

A vehicular network is such a complex system that is characterized by unpredictable vehicle movements, a dynamic topology, unstable communication connections, and frequent handover events. Computation task offloading and resource scheduling in vehicular network are a challenge, since an optimal solution should be aware of the network environment, understand the service requirements, and consider numerous other factors. Leveraging a machine learning approach in vehicular edge management is a promising paradigm for addressing the challenges mentioned above.

Various types of machine learning techniques have been applied in VEC, among which reinforcement learning is the most important. Reinforcement learning makes agents gain experience from their interactions with the environment and adjust action strategies along the learning process. This learning mode is suitable for dynamic road traffic states and complex vehicular networks. However, in large-scale networks handling massive amounts of state information, especially states represented by continuous values, the reinforcement learning approach cannot be directly implemented to solve the edge management problem. To address this issue, we can resort to deep Q reinforcement learning, which uses a Q-function as an approximator to capture complex interactions among various states and actions. Moreover, in the context of vehicular networks, some offloading actions could be chosen from a continuous space, such as wireless spectrum allocations and transmission power adjustments. To address the demands of this action space, edge service scheduling utilizes deep deterministic policy gradient learning, a branch of deep reinforcement learning that concurrently learns policy and value functions in a policy gradient learning process.

Many studies have applied machine learning to vehicular edge management. Some research focused on the relations of offloading decisions in the time dimension. Since data transmission and task execution are hard to complete instantaneously, previous actions will affect subsequent decisions through an extension of edge service states. The action dependence raises challenges in the optimization of current offloading strategies. To address them, Qi, Wang, et al. [74] designed a vehicular knowledge-driven offloading decision framework that scruples the future data dependence of the following generated tasks and helps obtain the optimal action strategies directly from the environment.

Another intersecting research issue of AI-empowered VEC is the adaptability of learning models in the context of complex vehicular networks. Considering potentially multiple optimization goals for offloading service management and that a single learning model can meet only part of the requirements, the incorporation of multiple models in the learning process is a promising approach. Sonmez, Tunca, Ozgovde and Ersoy [75] proposed a two-stage machine learning mechanism that consists of classification models in the first stage to improve the task completion success rate and regression models in the second stage to minimize edge service time costs. In [76], multi-model cooperation is adapted to become more flexible. To address the diversity and dynamics of the factors impacting edge service in vehicular networks, Chen, Liu, et al. introduced a meta-learning approach that adaptively selects the appropriate machine learning models and achieves the lowest offloading costs under different scenarios.

Running a learning process always consumes a great deal of computing resources, which will further aggravate the tension of edge service capabilities. To address this issue, a flexible, efficient, and lightweight learning mechanism is strongly needed. Research that has focused on this issue includes [77], where the authors aimed to reduce the learning complexity and processing costs. In the deep reinforcement learning–based offloading schemes proposed, Zhan, Luo, et al. avoided a large number of inefficient exploration attempts in the training process by deliberately adjusting the state and reward representations. Wang, Ning, et al. [78] presented an imitation learning–enabled online task scheduling scheme. In this scheme, the learning agents find optimal offloading strategies by solving an optimization problem with a few offline samples; near-optimal edge service performance is then achieved at a low learning cost.

5.5 A Case Study

In this section, we present two case studies to illustrate the vehicular task offloading mechanisms. The first one incorporates vehicle mobility into edge service manage-ment and proposes a predictive task offloading strategy [79]. The second one focuses on computation offloading in complex vehicular networks with multiple optional target servers and diverse data transmission modes, and it leverages AI technique to design optimal offloading schemes [80].

5.5.1 Predictive Task Offloading for Fast-Moving Vehicles

5.5.1.1 A System Model

We consider an MEC-empowered vehicular network, as illustrated in Fig. 5.2. Unin-terrupted traffic in a free flow state is running on a unidirectional road. Along the road are RSUs. The distance between two adjacent RSUs is L. The transmission range of each RSU is $L/2$. The road can be divided into several segments of length

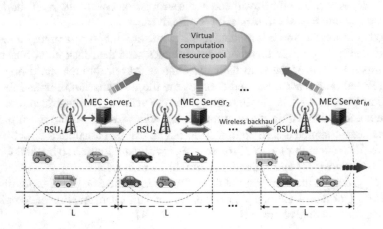

Fig. 5.2 An MEC-empowered vehicular network

L. Through the V2I communication mode, vehicles traveling on a given segment can only access the RSU located in the corresponding segment.

In the scenarios we studied, such as a temporarily deployed vehicular network, the RSUs communicate with each other through wireless backhauls. Each RSU is equipped with an MEC server with limited computational resources. To improve the transmission efficiency of the wireless backhauls, the task input file cannot be transmitted between the RSUs. Moreover, since the task output data size is small, the computation output can be transmitted between RSUs through wireless backhauls. All the vehicles move at a constant speed. The distribution of the vehicles on the road follows a Poisson distribution with density λ.

Each vehicle has a computation task. The task can be either carried out locally by the vehicular terminal or computed remotely on the MEC servers. The computation task is denoted as $T = \{c, d, t_{max}\}$, where c is the amount of the required computational resources, d is the size of the computation input file, and t_{max} is the delay tolerance of the task. We further classify the tasks into S types and present the tasks as $T_i = \{c_i, d_i, t_{i,max}\}$, $i \in S$. The vehicles can be correspondingly classified according to their computation task types into S types. The proportion of vehicles with a task of type i in the total number of vehicles on the road is ρ_i, where $i \in S$ and $\sum_{i=1}^{S} \rho_i = 1$.

5.5.1.2   Offloading with Optimal Predictive Transmission

There are two transmission modes for task offloading. One is through a direct V2I mode. In this mode, a vehicle can only offload its task to the MEC server equipped on the RSU that the vehicle can currently access. Considering that a vehicle travels down an expressway at high speed, if its computation task costs a relatively long time, the vehicle can pass by several RSUs during the task execution period. In this case, the output of the computation to be sent back to the vehicle needs to be transmitted from the MEC server that has accomplished the task to the remote RSU that the vehicle is newly accessing. The time overhead and transmission cost of the multi-hop relay seriously degrade the task transmission's effectiveness.

Another offloading mode is predictive V2V transmission, whose main framework is illustrated in Fig. 5.3. In this mode, the vehicles send their task input files to the MEC servers ahead of them, in their direction of travel, through multi-hop V2V relays. Based on the accurate prediction of the file transmission time and the task execution time, as well as the time spents for the vehicle traveling down the road, vehicle k can arrive within the communication area of RSU_n at the exact time its task has been completed. The computation output can be transmitted directly from RSU_n to the vehicle through V2I transmission without a multi-hop backhaul relay. Transmission costs for task offloading can thus be reduced.

Let $t_{i,v2v}$ denote the average time delay for the transmission of the input file of a task of type i through a one-hop V2V relay. The total time consumption of completing the task in this predictive mode is

$$t_{i,j} = y_j \cdot t_{i,v2v} + t_{i,upload} + t_{i,remote} + t_{i,download} \tag{5.1}$$

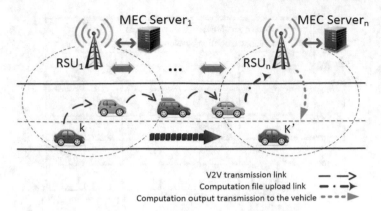

Fig. 5.3 Vehicle mobility-aware predictive task data transmission

where j is the number of hops the upload destination RSU is from the vehicle's current position, where $j > 1$ means the vehicles adopt predictive mode transmission. We define y_j as the number of V2V relay hops that are required to transmit the input file to an RSU j hops away. Furthermore, the total cost of this type of task offloading is

$$f_{i,j} = y_j \cdot f_{i,v2v} + f_{i,upload} + f_{i,remote} + f_{i,download} \tag{5.2}$$

where $1 < j \leq J_{i,\max}$.

To minimize the offloading cost of both data transmission and task execution while satisfying the latency constraints, the objective function of the optimal offloading schemes is

$$\min_{\{P_{i,j}\}} \sum_{i=1}^{S} \sum_{j=0}^{J_{i,\max}} \rho_i P_{i,j} f_{i,j} \tag{5.3}$$

$$\text{such that } t_{i,j} \leq t_{i,\max}, \quad i \in \{1, S\}, j \in \{0, J_{i,\max}\}$$

The objective function in (5.3) gives the average offloading costs of all types of vehicles when they choose offloading strategies $\{P_{i,j}\}$, where $\{P_{i,j}\}$ is the probability of a vehicle of type i choosing to offload its task to the MEC server j road segments away from its current position. To solve (5.3), we resort to a game approach to find the optimal offloading strategies of each type of vehicles. This game involves S players, where each player is a set of vehicles with the same type of tasks. We denote the vehicle set with tasks of type i as set i. The strategies of vehicle set i ($i = \{1, 2, \ldots, S\}$) are $\{P_{i,j}\}$. Vehicles in set i can choose to either execute tasks locally or offload them to MEC servers j hops away. The payoff for set i is the sum of the vehicles' offloading costs. Using a heuristic method in which each vehicle set adopt its best response action given the strategies of other vehicle sets, we can obtain a Nash equilibrium, which is the solution of (5.3).

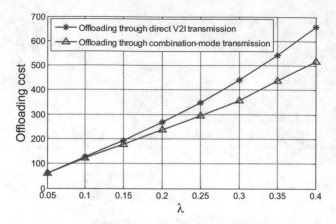

Fig. 5.4 Task offloading costs in terms of vehicles density

5.5.1.3   Performance Evaluation

In the simulation scenario, we consider 10 RSUs located along a four-lane one-way road. The vehicles are traveling at 120 km an hour. Their computation tasks are classified into five types, with the probabilities {0.05, 0.15, 0.3, 0.4, 0.1}, respectively. In addition, we set the computation resource requirement of each type of task at {7, 13, 27, 33, 48} units, respectively.

Figure 5.4 shows the computation offloading costs with different densities of vehicles on the road. We compare the performance of our proposed predictive offloading scheme with the V2I direct transmission scheme. It can be seen that the predictive scheme greatly reduces the cost when the road has high vehicle density. In the case of high traffic density, long task execution times on the MEC servers lead to more RSUs that the vehicles have traveled past. Due to the transmission cost of the wireless backhaul between RSUs, the total costs of the direct V2I scheme rise quickly with an increase in the density λ. However, in the predictive scheme, part of the transmission is offloaded to the V2V relay, which has a lower cost compared with wireless backhaul transmission. Thus, computation offloading costs can be saved.

It is worth noting that the performance improvement brought about by predictive offloading is based on the accurate prediction of vehicle mobility. With the development of AI technology, the prediction of vehicle mobility patterns has become much more accurate, especially on highways that have stable traffic flows. Thus, this proposed predictive scheme is promising and effective in practical applications.

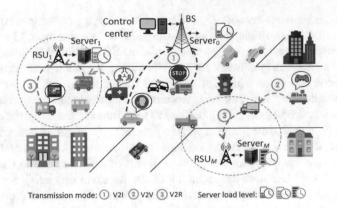

Transmission mode: ① V2I ② V2V ③ V2R   Server load level: 🕐 🕐 🕐

Fig. 5.5 Task offloading in an MEC-enabled vehicular network

5.5.2 Deep Q-Learning for Vehicular Computation Offloading

5.5.2.1 A System Model

In this case study, we consider an MEC-enabled vehicular network in an urban area, as illustrated in Fig. 5.5. Various types of computation tasks are generated in the traveling vehicles. We classify these tasks into G types. A task is described in four terms, as $\kappa_i = \{f_i, g_i, t_i^{\max}, \varsigma_i\}$, $i \in G$, where f_i and g_i are the size of the task input data and the amount of required computation, respectively, and $t_i^{\max}$ is the maximum delay tolerance of task κ_i. The offloading system receives utility $\varsigma_i \Delta t$ upon completion of task κ_i, where Δt is the time saved in accomplishing κ_i compared to $t_i^{\max}$. The probability of a task belonging to type i is denoted as β_i, with $\sum_{i \in G} \beta_i = 1$.

The urban area is covered by a heterogeneous wireless network that consists of a cellular network BS, M RSUs, and mobile vehicles. Compared to a BS that has seamless coverage and a high data transmission cost, RSUs provide spotty coverage but inexpensive access service. The costs for using a unit of the spectrum of the cellular network and the spectrum belonging to the vehicular network per unit of time are c_c and c_v, respectively.

The BS is equipped with an MEC server, denoted as Serv_0, through wired connections. In addition, each RSU hosts an MEC server. These servers are denoted Serv_1, $\text{Serv}_2, \ldots, \text{Serv}_M$, respectively. The MEC servers receive data from their attached BS or RSUs directly. Let $\{W_0, W_1, W_2, \ldots, W_M\}$ denote the computing capacities of these servers. Each MEC server is modeled as a queuing network where the input is the offloading task. The tasks that arrive are first cached on an MEC server and then served according to a first-come, first-served policy. A server utilizes all of its computing resources to execute the currently served task. The cost for a task to use a computing resource per unit of time is c_x.

In the heterogeneous network formed by the overlapping coverage of the BS and the RSUs, vehicles can offload their tasks to the MEC servers through multiple modes. The task file transmission between a vehicle and the BS is called V2I. When a vehicle turns to the LTE-Vehicle network for task offloading, the file can be transmitted to an MEC server in a mode with joint V2V–V2R transmission.

The task offloading scheduling and resource management are considered to operate in a discrete time model with fixed length time frames. The length of a frame is denoted as τ. In each time frame, a vehicle generates a computing task with probability P_g. For each generated task, its offloading can only adopt a single transmission mode. Since the topology can change in different time frames due to the mobility of the vehicles, to facilitate the modeling of the dynamic offloading service relations, we split the road into E segments. The position of a vehicle on the road is denoted by the index of the segment e, where $1 \leq e \leq E$. All the vehicles have fixed transmission power for a given transmission mode, that is, $P_{tx,b}$ in V2I mode and $P_{tx,v}$ in the V2R and V2V modes. To receive a task file from a V2I mode vehicle, the signal-to-noise-plus-interference ratio (SINR) at the BS is presented as $\gamma_{v,b}$. In addition, when vehicles choose V2R or V2V communication, the SINR at receiver r is $\gamma_{v,r}$.

5.5.2.2 Optimal Offloading Scheme in a Deep Q-Learning Approach

We next formulate an optimal offloading problem and propose deep Q-learning–based joint MEC server selection and offloading mode determination schemes. In a given time frame, for a vehicle located on road segment e and generating task κ_i, we use $x_{i,e} = 1$ to indicate the task offloading to Serv_0 through V2I. Similarly, we use $y_{i,e,m} = 1$ and $z_{i,e,m} = 1$ to indicate the task offloading to Serv_m in the V2R and joint V2V–V2R modes, respectively. Otherwise, these indicators are set to zero. The proposed optimal task offloading problem, which maximizes the utility of the offloading system under task delay constraints, is formulated as follows:

$$
\begin{aligned}
\max_{\{x,y,z\}} U = \sum_{l=1}^{\infty} \sum_{j=1}^{n} \sum_{i=1}^{G} \beta_i \big(& \varsigma_i (t_i^{\max} - t_{i,e_j,l}^{\text{total}}) - x_{i,e_j}^l (q_c c_c f_i \\
& /R_{v,b,e_j} + g_i c_x / W_0) - y_{i,e_j,m}^l (q_v c_v f_i / R_{v,r,e_j} \\
& + g_i c_x / W_m) - z_{i,e_j,m}^l \big(\sum_{h=1}^{H_{e_j}} q_v c_v f_i / R_{v,j,e_j} + g_i c_x / W_m \big) \big) \\
\text{such that} \quad & C1 : x_{i,e_j}^l = \{0,1\}, \ y_{i,e_j,m}^l = \{0,1\}, \ z_{i,e_j,m}^l = \{0,1\} \\
& C2 : x_{i,e_j}^l y_{i,e_j,m}^l = x_{i,e_j}^l z_{i,e_j,m}^l = y_{i,e_j,m}^l z_{i,e_j,m'}^l = 0 \\
& C3 : x_{i,e_j}^l + y_{i,e_j,m}^l + z_{i,e_j,m}^l = 1 \\
& C4 : t_{i,e_j}^{\text{total}} \leqslant t_i^{\max}, \quad i \in \kappa, \quad m, m' \in M
\end{aligned}
\tag{5.4}
$$

where n is the number of tasks generated in a time frame; e_j is the road segment index of vehicle j's location; H_{e_j} is the number of transmission hops; q_c and q_v

are the amount of spectrum resources allocated for each task file offloading through the cellular and vehicular networks, respectively; R_{v,b,e_j} is the transmission rate of offloading the task file from the vehicle in road segment e_j to the BS, which can be written $R_{v,b,e_j} = q_c \log(1 + \gamma_{v,b})$; and R_{v,r,e_j} and R_{v,j,e_j} can be calculated similarly, based on the allocated spectrum q_v and the SINR $\gamma_{v,r}$. Constraint C1 indicates that, for any offloading mode, either a choice is made or not, and C2 and C3 ensure that each task should select only a single offloading mode.

Since the current serving state of a server can affect the time costs of the following tasks, we can formulate (5.4) as a Markov decision process. The state of the offloading system in time frame l is defined as $S^l = (s_0^l, s_1^l, \ldots, s_M^l)$, where s_0^l is the total computation required by the tasks queuing in Serv_0 in frame l. Similarly, $s_1^l, \ldots, s_M^l$ denote the required computation of the tasks queuing in $\text{Serv}_1, \text{Serv}_2, \ldots, \text{Serv}_M$ in time frame l, respectively. The actions taken by the control center in frame l can be shown to be $a^l = (X^l, Y^l, Z^l)$, where $X^l = \{x_{i,e}^l\}$, $Y^l = \{y_{i,e,m}^l\}$ and $Z^l = \{z_{i,e,m}^l\}$ are the sets of task offloading strategies with various transmission modes and offloading targets for the generated tasks in frame l, respectively.

To derive the optimal offloading strategy π^*, we turn to reinforcement learning technology. The previous Markov decision process is turned into a reinforcement learning problem. The optimal value of the Q-function is

$$Q^*(S^l, a^l) = \mathrm{E}_{S^{l+1}}[U^l + \eta \max_{a^{l+1}} Q^*(S^{l+1}, a^{l+1})|S^l, a^l] \tag{5.5}$$

where the maximum-utility as well as optimal offloading strategies can be derived by value and strategy iteration. A classical algorithm of reinforcement learning technologies, Q-learning can be used in modifying the iterations. In each iteration, the value of the Q-function in the learning process is updated as

$$Q(S^l, a^l) \leftarrow Q(S^l, a^l) + \alpha[U^l + \eta \max_{a^{l+1}} Q^*(S^{l+1}, a^{l+1}) - Q(S^l, a^l)] \tag{5.6}$$

where α is the learning rate.

Moreover, the states of the offloading system consist of the amount of computation queuing required in the MEC servers, a continuous value. We thus transform the Q-function into a function approximator and choose a multilayered neural network as a nonlinear approximator that can capture complex interactions among various states and actions. Based on the Q-function estimation, we utilize deep Q-learning technology to obtain the optimal offloading strategies π^*. With the help of the Q-network, the Q-function can be estimated as $Q(S^l, a^l) \approx Q'(S^l, a^l; \theta)$, where θ is the set of network parameters. The Q values are trained to converge to real Q values over iterations. Based on the Q values, the optimal offloading strategies in each state are derived from the actions that lead to maximum utility. The action chosen in frame l can now be written as $a^{l*} = \arg\max_{a^l} Q'(S^l, a^l; \theta)$. During Q-learning updates, a batch of stored experiences drawn randomly from the replay memory are used as samples in training the Q-network's parameters. The goal of the training is to minimize the difference between $Q(S^l, a^l)$ and $Q'(S^l, a^l; \theta)$. The loss function is given as

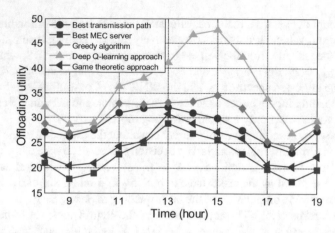

Fig. 5.6 Average utilities under different offloading schemes

$$Loss(\theta^l) = E[\frac{1}{2}(Q_{tar}^l - Q'(S^l, a^l; \theta^l))^2] \qquad (5.7)$$

We deploy a gradient descent approach to modify θ. The gradient derived through differentiating $Loss(\theta^l)$ is calculated as

$$\nabla_{\theta^l} Loss(\theta^l) = E[\nabla_{\theta^l} Q'(S^l, a^l; \theta^l)(Q'(S^l, a^l; \theta^l) - Q_{tar}^l)] \qquad (5.8)$$

Then, θ^l is updated according to $\theta^l \leftarrow \theta^l - \varpi \nabla_{\theta^l} Loss(\theta^l)$ in time frame l, where ϖ is a scalar step size.

5.5.2.3 Performance Evaluation

We evaluate the performance of the proposed task offloading schemes based on real traffic data, which consist of 1.4 billion GPS traces of more than 14,000 taxis recorded during 21 day in a city. We consider a scenario with one BS and five RSUs on each selected road. We set a computing capacity $W_0 = 1,000$ units, and the capacities of the MEC servers equipped on the RSUs are randomly selected from the range of [100, 200] units.

Figure 5.6 shows the impact of road traffic on the average utility of a task under different offloading schemes. Our proposed deep Q-learning scheme clearly yields higher offloading utility compared to other schemes, especially in the non-rush period from 12:00 to 16:00. This is because our scheme jointly considers transmission efficiency and the load states of the MEC servers. However, the offloading scheme that chooses the target server according to the vehicle's best transmission path and the scheme that selects the MEC server according to the server state only take one factor into account. The ignored factor could seriously affect the offloading efficiency.

In the game-theoretic approach, the vehicles traveling on a road segment act as players that compete for task offloading services to gain higher utility. Since each vehicle independently determines its offloading strategy from the perspective of its own interests and ignores cooperation with other vehicles, system performance worsens. In the greedy algorithm, each vehicle chooses its offloading strategy in a distributed manner. Although the greedy algorithm jointly optimizes the file trans-mission path and MEC server selection in the current frame, it ignores the follow-up effects between consecutive time frames. In contrast, our proposed learning scheme considers both of these effects in the design of offloading strategies, leading to better performance.

Chapter 6
Mobile Edge Computing for UAVs

Abstract This chapter studies mobile edge computing (MEC) networks assisted by unmanned aerial vehicles (UAVs). According to the application scenarios, we consider three roles for UAVs in MEC networks: exploiting MEC computing capabilities, serving as a computing server, and serving as a relay for computation offloading. Furthermore, the details for resource allocation and optimization are presented in the three scenarios of UAV-assisted MEC networks. In addition, we focus on the situation in which a UAV not only functions as an MEC server to inspect turbines on a wind farm, but also performs task computation. To facilitate wide applications of UAV-assisted MEC in practice, this chapter highlights the main implementation issues of UAV-assisted MEC, including optimal UAV deployment, wind models, and joint trajectory–computation performance optimization.

6.1 Unmanned Aerial Vehicle–Assisted Mobile Edge Computing (MEC) Networks

MEC has emerged as a promising solution to enable resource-limited mobile devices to execute real-time applications (e.g., face recognition, augmented reality, unmanned driving) [81]. With the deployment of MEC servers—such as base stations (BSs) and access points—at the network edges, mobile users can offload computation-intensive and latency-critical tasks to the network edges for computing, instead of to the central cloud, to improve the computation performance of mobile users in a cost-effective and energy-saving manner. However, terrestrial MEC networks are not reliably established in some scenarios, such as in disasters, on battle fields, and in emergency areas. Recently, MEC assisted by unmanned aerial vehicles (UAVs) has drawn significant research interest because of the advantages it offers, such as fully controllable mobility, flexible deployment, and strong line-of-sight channels with ground devices [82]. Therefore, UAV-assisted MEC can be flexibly deployed in scenarios where terrestrial MEC networks might not be convenient.

As shown in Fig. 6.1, under an emergency scenario, terrestrial MEC infrastructures could be destroyed in a disaster, leaving many rescue tasks unable to be computed or executed. UAVs mounted with edge servers could be dispatched to compute the

© The Author(s) 2022
Y. Zhang, *Mobile Edge Computing*, Simula SpringerBriefs on Computing 9,
https://doi.org/10.1007/978-3-030-83944-4_6

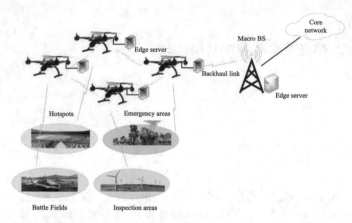

Fig. 6.1 A UAV-assisted MEC network framework

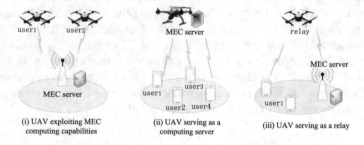

Fig. 6.2 Three UAV-assisted MEC network scenarios

rescue tasks in time. It is sometimes difficult to establish terrestrial MEC networks to compute inspection tasks, such as for turbines at offshore wind farms or power lines in a smart grid, due to a harsh environment. In this case, a UAV-assisted MEC network can play a very important role. In hotspots, the high volumes of computation tasks offloaded from millions of mobile users can exhaust the computation resources of edge servers, which leads to increased processing latency that decreases the user's quality of experience. The assistance of UAVs can improve the user's quality of experience.

Recent research has focused on the advances of employing UAV-assisted MEC to help ground mobile users [83–85, 88, 90, 95, 96]. As shown in Fig. 6.2, according to the application scenarios, UAVs can play different roles in MEC networks [83], as follows.

- *Exploiting MEC computing capabilities:* When the UAV has limited computation capability and needs to execute computation-intensive tasks, it can function as a user to offload tasks to the terrestrial MEC server for computing.

- *Serving as a computing server:* When terrestrial MEC networks are not reliably established, the UAV functions as an MEC server to help ground mobile users perform tasks computation.
- *Serving as a relay for computation offloading:* When the UAV is not equipped with an MEC server and the offloading link between the mobile user and the terrestrial BS with the MEC server is poor, the UAV works as a relay to assist the mobile user offload tasks to the terrestrial BS.

6.2   Joint Trajectory and Resource Optimization in UAV-Assisted MEC Networks

Resource allocation and optimization are important for improving computation performance while realizing the economical operation of UAV-assisted MEC networks. Unlike resource allocation in conventional terrestrial MEC networks, resource allocation in UAV-assisted MEC networks must consider the resources to be optimized in computation task offloading, local computation, and UAV flight [83]. The details for the resource allocation and optimization are presented in the following for the three UAV-assisted MEC network scenarios.

6.2.1   Resource Allocation and Optimization in the Scenario of a UAV Exploiting MEC Computing Capabilities

In the scenario of exploiting MEC computing capabilities, UAVs can exploit a partial offloading mode or a binary offloading mode to offload computation-intensive tasks to the terrestrial MEC server for computation. For computation task offloading, the multiple access techniques used for terrestrial MEC networks can also be applied to the UAVs' computation task offloading procedure. Multiple access techniques can be classified into two categories: orthogonal multiple access (OMA) and non-orthogonal multiple access (NOMA). The typical OMA techniques used in UAV-assisted MEC are time division multiple access (TDMA) and orthogonal frequency division multiple access (OFDMA). NOMA can improve user connectivity and spectral efficiency compared with OMA. In the UAVs' computation task offloading procedure, the terrestrial MEC server allocates the communication resources (communication bandwidth, offloading time, offloading power, etc.) for task offloading. In local computation, the terrestrial MEC server allocates and optimizes the CPU frequency for the computation of the UAVs' tasks. For the UAVs' flight, their trajectories are optimized under speed and acceleration constraints.

Moreover, different objectives can be achieved through resource allocation and optimization in the scenario of a UAV exploiting MEC computing capabilities, such as energy consumption minimization, completion time minimization, and utility maximization.

6.2.1.1 Energy Consumption Minimization

The energy consumption in the case of a UAV exploiting MEC computing capabilities results from the UAV's task offloading process, the local computing process, and the UAV's flight. In [84], N. Motlagh et al. studied the offloading of video data processing to an MEC node compared to the local processing of video data on UAVs. Moreover, a testbed was developed to demonstrate the efficiency of the MEC-based offloading approach in saving the scarce energy of UAVs. In [85], M. Hua et al. studied energy consumption minimization for computation task offloading from multiple UAVs to a terrestrial BS, comparing the performance impacts of different the access schemes of multiple UAVs.

6.2.1.2 Completion Time Minimization

In [86], X. Cao et al. considered an MEC system with a single UAV and a set of ground BSs with MEC functionality. The UAV offloads computation tasks to ground BSs for remote execution. The computation tasks can be arbitrarily partitioned into smaller subtasks that can be offloaded to different ground BSs. The mission completion time of these subtasks is discretized into N time slots, whose number is minimized by jointly optimizing the UAV trajectory and computation offloading, subject to the UAV's maximum speed constraint and the computation capacity constraint of the ground BSs.

6.2.1.3 Utility Maximization

In [87], the authors defined the utility of UAV task offloading by considering energy consumption, time delay, and computation cost. The best possible trade-off between energy consumption, time delay, and computation cost can be achieved by maximizing a global utility function.

6.2.2 Resource Allocation and Optimization in the Scenario of a UAV Serving as a Computing Server

In the scenario of a UAV serving as a computing server, terrestrial users exploit a partial offloading mode or a binary offloading mode to offload computation-intensive tasks to the UAV for computation. The multiple access techniques of

TDMA, OFDMA, and NOMA can be used to offload terrestrial user tasks. In the offloading procedure, the UAV allocates the communication resources (communication bandwidth, offloading time, offloading power, etc.) to terrestrial users for task offloading. For the UAV flight and computing process, the resources to be optimized are the CPU frequency, the trajectory, the flight speed, and the acceleration velocity of the UAV. There are two computing techniques based on the UAV CPU frequency. When the UAV computation circuit has a fixed CPU frequency, the local computation is performed at a constant rate. When the UAV adopts a dynamic voltage and frequency scaling technique, the CPU frequency can be adjusted based on the scale of the computation task. Different optimization objectives can be achieved through resource allocation when the UAV serves as a computing server, such as energy consumption minimization, computation bit maximization, completion time minimization, and computation efficiency maximization.

6.2.2.1 Energy Consumption Minimization

The energy consumption in the case of a UAV serving as a computing server results from the local computing process, the task offloading process, and the UAV's flight. In particular, the energy consumed in the local computing process is determined by the CPU frequency. In the offloading process, the energy consumed depends on the transmission power and the offloading time. In the UAV's flight, the energy consumed should consider the UAV's speed, acceleration velocity, and flight time. Joint trajectory and resource allocation was studied in [88] to minimize total mobile energy consumption while satisfying the quality of service requirements of the offloaded mobile application. A UAV-assisted MEC network with stochastic computation tasks was investigated in [89], where the average weighted energy consumption of the smart devices and the UAV is minimized subject to constraints on the UAV's computation offloading, resource allocation, and flight trajectory scheduling.

6.2.2.2 Computation Bit Maximization

The total number of computation bits is the sum of the number executed in the local computation and the number executed in the offloading. The maximization of the number of computation bits was studied in a UAV-assisted MEC wireless powered system under both partial and binary computation offloading modes in [90], subject to a causal energy harvesting constraint and a UAV speed constraint. The weighted-sum completed task input bits of users was maximized in [91] under task and time allocation, information causality, energy causality, and the UAV's trajectory constraints. The optimization problem was solved by jointly optimizing the task and time allocation, as well as the UAV's transmission power and trajectory.

6.2.2.3   Completion Time Minimization

The completion time, in the case of a UAV serving as a computing server, is another fundamental optimization objective, especially for time-sensitive applications. In the binary computation mode, the completion time can be determined by the local computation time or the offloading time. In the partial computation mode, the completion time depends on both the local computation time and the offloading time. The minimization of the task completion time is studied in [92] while assuming the condition of a minimum number of computation bits.

6.2.2.4   Computation Efficiency Maximization

Different from the individual optimization of computation latency, energy consumption, and the number of computation bits, computation efficiency is defined as the ratio of the total number of computation bits to the total energy consumption, to achieve a good trade-off between the number of computation bits and energy consumption. A computation efficiency maximization problem was formulated in a multi-UAV-assisted MEC system in [93], where, based on the partial computation offloading mode, user association, the allocation of CPU cycle frequencies, power and spectrum resources, as well as UAV trajectory scheduling are jointly optimized.

6.2.3   Resource Allocation and Optimization in the Scenario of a UAV Serving as a Relay for Computation Offloading

In the scenario of a UAV serving as a relay for computation offloading, users deployed at the cell edge offload computation-intensive tasks to a terrestrial MEC server via UAV relay for computing. In the offloading procedure, the terrestrial MEC server allocates the communication resources (e.g., communication bandwidth, offloading time, relaying power) to the users and the UAV for task offloading and relaying. In the local computation, the terrestrial MEC server allocates and optimizes the CPU frequency for users' task computations. In the UAV flight process, the UAC's trajectory is optimized under speed and acceleration constraints. The different optimization objectives can be achieved through resource allocation where the UAV serves as a relay, such as in user latency minimization, energy consumption minimization, and minimum throughput maximization.

6.2.3.1 User Latency Minimization

In [94], J. Lyu et al. proposed utilizing UAVs as computing nodes as well as relay nodes to improve the average user latency. They formulated the optimization problem with the objective of minimizing the average latency of all users.

6.2.3.2 Energy Consumption Minimization

In [95], J. Lyu studied a UAV-assisted MEC architecture where a UAV roamed around the area, serving as a server to compute the tasks of mobile users or acting as a relay to offload computation tasks to an access point. The problem of minimizing the weighted-sum energy consumption of the UAV and the mobile users was formulated subject to task constraints, information causality constraints, bandwidth allocation constraints, and UAV trajectory constraints.

6.2.3.3 Minimum Throughput Maximization

In [96], a UAV was leveraged as a relay for offloading the computation tasks of mobile users to a terrestrial BS. Furthermore, a resource optimization problem for maximizing the minimum throughput of mobile devices was formulated. The results demonstrate that computation performance can be significantly improved by optimizing UAV trajectories.

6.3 Case Study: UAV Deployment and Resource Optimization for MEC at a Wind Farm

Most recent research has focused on advances employing UAV-assisted MEC to help ground mobile users in task computation or relaying in a hotspot scenario. Nevertheless, UAV-assisted MEC has seldom been studied in a harsh environment (at a wind farm, in rough seas, etc.). In this section, we consider the situation in which a UAV functions as an MEC server to inspect the turbines at a wind farm and performs task computation. Wind power is a clean and widely deployed alternative for reducing dependence on fossil fuel power. Under this trend, a large number of turbines are being installed at wind farms. The deployment of UAVs for the automated inspection of the turbines and related task computation is a promising method to reduce costs and improve inspection and computation efficiency. Different from UAV-assisted MEC in a hotspot scenario, the random realization of extreme weather conditions at a wind farm impacts the flight speed and range of UAVs [99]. For example, UAVs can crash if the wind speed at a wind farm is over their maximum wind speed resistance. Therefore, the influence of wind speed and wind direction is important to consider in UAV placement and routing at a wind farm. In this section, we study the optimal deployment of UAVs to inspect all turbines at a wind farm. The joint trajectory–resource optimization for inspection and task computation will be discussed in the following sections.

6.3.1 UAV Deployment for MEC at a Wind Farm

6.3.1.1 A Wind Model

Figure 6.3 shows the deployment of multiple UAVs to monitor the condition of turbines. The total number of turbines at the wind farm is T. The coordinates of the kth turbine are $\mathbf{q}_k = [x_k, y_k]$. When UAVs fly to inspect a wind turbine, they face two wind conditions: tail wind and head wind. The wind velocity is $\mathbf{w} = [w^x, w^y]$, and the wind speed is $w_s = \|\mathbf{w}\|_2$. In the polar coordinate system, the wind direction is θ_w^{pol}, which can be calculated by $\theta_w^{pol} = arc \tan \frac{w^y}{w^x}$. In meteorological measurements, the wind direction is θ_w^{met}. Since the phase goes clockwise in meteorological measurements, however, the phase in the polar coordinate system is represented in the counterclockwise direction. Therefore, the relation between wind direction in the polar coordinate system and in meteorological measurement is denoted as

$$\theta_w^{pol} = \frac{3\pi}{2} - \theta_w^{met} \tag{6.1}$$

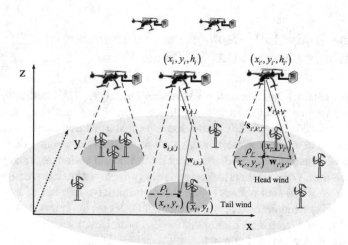

Fig. 6.3 The deployment of multiple UAVs for automated inspection at a wind farm

6.3.1.2   A UAV Model

At the wind farm, each UAV inspects the turbines that are assigned to it. Assume that the UAVs all fly at the same altitude, so the z-axis can be ignored. The ith UAV will be placed at $\mathbf{q}_i = [x_i, y_i]$, and the set of turbines assigned to the ith UAV is denoted by $\mathcal{N}_i$. When a UAV flies to inspect a wind turbine, the condition of the wind is to be considered in the decision making of the UAV. The velocity of the ith UAV flying from turbine k to turbine l is $\mathbf{v}_{i,k,l} = [v^x_{i,k,l}, v^y_{i,k,l}]$, which is the UAV's initial velocity. The resultant velocity of the ith UAV is $\mathbf{s}_{i,k,l} = [s^x_{i,k,l}, s^y_{i,k,l}]$, which is the velocity influenced by the wind. The relation between the UAV velocity, the wind, and the resultant velocity can therefore be expressed as

$$\mathbf{s}_{i,k,l} = \mathbf{v}_{i,k,l} + \mathbf{w} \tag{6.2}$$

where $\|\mathbf{v}_{i,k,l}\|_2$ and $\|\mathbf{s}_{i,k,l}\|_2$ are the airspeed and ground speed of the UAV, respectively, both limited by the maximum speed limit of u^{max}_i. In particular, when the UAV faces a tail wind, the ground speed is limited to u^{max}_i. Then, for the headwind case, the airspeed is limited to u^{max}_i.

The time for the UAV to fly from turbine k to turbine l can be calculated as

$$t_{i,k,l} = \frac{\|\mathbf{q}_l - \mathbf{q}_k\|_2}{\|\mathbf{s}_{i,k,l}\|_2} \tag{6.3}$$

The maximum flight time for the UAV is t^{max}_i, which represents an upper limit of the total flight time. The UAV's flight range under the wind condition can be expressed as

$$B^{\mathbf{w}}_i(\rho_i) = \{x, y \in: \|\mathbf{r}\|_2 \leq \rho_i\} \tag{6.4}$$

where $\mathbf{r} = [x - x_r, y - y_r]$, ρ_i is the UAV's actual flight distance, which can be calculated by $\rho_i = \frac{u^{max}_i t^{max}_i}{2}$. The UAVs' flight range is regarded as a circle with x_r and y_r as the center of the circle, which can be calculated as $x_r = x_i + w^x t^{max}_i$ and $y_r = y_i + w^y t^{max}_i$, respectively. Since the UAV's flight range is influenced by different wind conditions, it is the intersection of the flight range under different wind conditions: $Z_i = \cap_{\mathbf{w}} B^{\mathbf{w}}_i(\rho_i)$.

6.3.1.3   Deployment of Multiple UAVs at a Wind Farm

Since a single UAV has a limited operation time and battery, it is challenging to serve a large number of mobiles users in a geographical coverage area. Compared with a single UAV, a collaboration of multiple UAVs can expand the coverage area and support more computation tasks within a shorter time, which can remarkably boost the applications of UAV-assisted MEC in emergency and complicated scenarios. Therefore, it is important to design a multiple-UAV deployment scheme before task computation, to reduce the computation cost and improve computation performance. Several recent works have focused on utilizing multiple UAVs and placing them for an optimal topology [97, 98]. An efficient deployment of multiple UAVs that provides coverage for ground mobile devices was analyzed in [97], where the three-dimensional locations of the UAVs were optimized to maximize the total coverage area. Furthermore, the minimum number of UAVs to guarantee a target coverage probability for a given geographical area was determined. A polynomial-time algorithm with the placement of successive UAVs was designed in [98] to minimize the number of UAVs needed to provide wireless coverage for a group of distributed ground terminals. However, recent works seldom study the deployment of UAVs for task computation in harsh environments, such as in rough seas or at a wind farm. We address the deployment problem of UAVs at a wind farm by considering wind, since it can have a considerable influence on the flight range and speed of UAVs.

Initially, N candidates UAVs are placed at the wind farm. The matrixes $\mathbf{A} = [a_i]_{1 \times N}$, $\mathbf{B} = [b_{i,k}]_{N \times T}$ and $C = [c_{i,j}]_{N \times N}$ denote the states of candidate UAVs, the association between UAVs and turbines, and the communication link between UAVs, respectively, where $a_i = 1$ indicates that the ith candidate UAV needs to be positioned at the wind farm, and $a_i = 0$ indicates it needs to be removed. Additionally, if $b_{i,k} = 1$, turbine k is assigned to the ith UAV; otherwise, $b_{i,k} = 0$. When determining the topology of the UAVs, we need to ensure that they can maintain communication links with each other so that collisions can be prevented. Thus, $c_{i,j} = 1$ indicates that the ith and jth UAVs have a communication link; otherwise, $c_{i,j} = 0$. The objective function is to minimize the number of UAVs that need to be placed at the wind farm. Thus, the deployment problem, considering the influence of wind, can be formulated as

$$\min_{\mathbf{A},\mathbf{B},C,x_i,y_i} \sum_{i=1}^{N} a_i$$

such that

$$C1 : a_i, b_{i,k}, c_{i,j} \in \{0, 1\}$$
$$C2 : \sum_{i=1}^{N} b_{i,k} \leq 1, \forall k, \sum_{k=1}^{T} b_{i,k} \leq p, \forall i \qquad (6.5)$$
$$C3 : [x_k, y_k] \in Z_i, [x_i, y_i] \in \{[x_k, y_k]\}, \forall b_{i,k} = 1$$
$$C4 : \sum_{j=1,i \neq j}^{N} c_{i,j} \geq 1, \sqrt{(x_i - x_j)^2 + (y_i - y_j)^2} \leq d, \forall i, j$$

where constraint C1 indicates that a_i, $b_{i,k}$, and $c_{i,j}$ are binary variables; C2 indicates that each turbine can only be assigned to one UAV, and each UAV can be associated with up to p turbines, respectively; C3 ensures that the location $[x_k, y_k]$ of turbine k assigned to the ith UAV must be in the flight range of the ith UAV, and each UAV should be placed inside the turbines assigned to it; and constraint C4 guarantees that the minimum number of communication links each UAV must have and the distance between any two UAVs should be lower than d. The formulation in (6.5) is a mixed integer linear programming problem that cannot be solved directly, because it contains binary parameters. In this case, if the dimension of the problem increases, the problem will become NP-hard. Thus, a heuristic algorithm is designed to solve problem (6.5).

For the deployment of multiple UAVs, the flight range is an important parameter that can be influenced by the wind conditions at the wind farm. However, wind conditions in the future are unknown to UAVs, and different UAVs have different maximum wind speed resistance levels. Since the deployment is used to determine the required number of UAVs and their placement, it can be considered the planning stage and can use historic wind data. We choose the Walney Wind Farms, a group of offshore wind farms in the United Kingdom. The wind data are obtained from the Centre for Environmental Data Analysis [103], which provides hourly average wind velocities and directions. Given the wind velocity and directions, a UAV's flight range can be obtained from (6.4).

Then, with a known flight range, the topology of the UAVs at the wind farm can be designed. Initially, T UAVs are placed at the wind farm so that each turbine has one UAV assigned to it. The turbines inside a UAV's flight range are assigned to it. After initialization, some UAVs might have to associate with more than p turbines. To eliminate redundant connections between the UAV and the turbines, the distances between the UAV and its assigned turbines are sorted in decreasing order and the turbines associated with the UAV are then reassigned accordingly.

After this procedure is completed, some turbines might be assigned to more than one UAV. In addition, the current number of UAVs placed has not been minimized. To address these issues, the UAVs are sorted based on the number of turbines intersecting with other UAVs in decreasing order. Then, if any turbine associated with the ith UAV can also be served by another UAV, the ith UAV can be removed; otherwise, the connection between the UAV and the turbine is deleted based on the distance.

6.3.1.4 Performance Analysis

The performance of the deployment strategy is evaluated based on a real-world dataset for the Walney Wind Farms in the United Kingdom. The data are from the Centre for Environmental Data Analysis [103] and Kingfisher Information Service – Offshore Renewable & Cable Awareness [104]. The UAV is an AscTec Falcon 8 [105], which has a maximum wind speed resistance of 15 m per second. The maximum distance of the communication between UAVs is set to 5 km. Each UAV can be assigned to inspect up to five turbines. To clearly present the deployment, we pick 47 of the 189 turbines in the dataset. Figure 6.4 shows the final deployment results, where 17 UAVs are required to cover all the wind turbines at the wind farm. Additionally, all the turbines are assigned to the UAVs, and each UAV serves no more than five turbines.

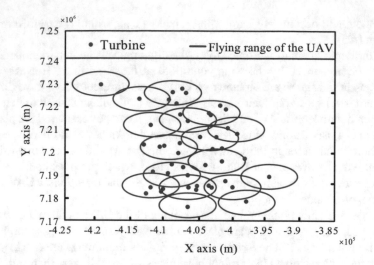

Fig. 6.4 UAV deployment

6.3.2 Joint Trajectory and Resource Optimization of UAV-Aided MEC at a Wind Farm

Different from the joint trajectory–resource optimization in UAV-assisted MEC for ground mobile users, in this section we study how to route UAVs to inspect wind turbines and execute computation tasks at a wind farm. For a given topology in Fig. 6.3, a UAV and its assigned turbines can be represented in a graph, denoted as $\mathcal{G}_i = \{\mathcal{N}_i, \varepsilon_i\}$, where $\mathcal{N}_i$ represents the nodes in the graph, which denote the set of turbines assigned to the ith UAV, and ε_i is the set of edges that connect each turbine. After the detection of turbine k, the UAV computes the detection tasks while flying to turbine l. The detection tasks must be completely processed before the UAV arrives at turbine l. With the graph structure, we can create adjacency matrixes denoted by $\mathbf{D}_i$ and $\mathbf{E}_i^m$. The value of the kth column and lth row in $\mathbf{D}_i$ is $t_{i,k,l}$, which represents the flight time from turbine k to turbine l. The value of the kth column and lth row in $\mathbf{E}_i^m$ is $e_{i,k,l}$, which represents the energy consumption of the UAV flying from turbine k to turbine l. The detection task of turbine k is denoted by $I_k = \{d_k, c_k\}$, where d_k denotes the task size and c_k is the required number of CPU cycles to compute one bit of task. Let $f_{i,k}$ denote the computation resource that the UAV allocates to turbine k. The computation time of the detection task for turbine k can be expressed as $t_{i,k} = \frac{c_k d_k}{f_{i,k}}$, where $t_{i,k} \leq t_{i,k,l}$. The energy consumption of the detection task computation for turbine k can be expressed as $e_{i,k} = \gamma_c c_k d_k \big(f_{i,k}\big)^2$.

Denote the number of required routes to inspect the turbines as M. We introduce the routing matrix $\mathbf{U}_i^m = [U_{i,k,l}^m]_{|\mathcal{N}_i| \times |\mathcal{N}_i|}$ to denote the mth route for UAV i. Specifically, $U_{i,k,l}^m = 1$ indicates that the UAV chooses to fly from turbine k to turbine l; otherwise, $U_{i,k,l}^m = 0$. Our objective is to minimize the energy consumption of the ith

UAV for turbine detection and task computation. The joint trajectory–computation resource optimization problem can be formulated as

$$\min_{\substack{M, U_i^m, E_i^m, f_{i,k}, \\ v_{i,k,l}, s_{i,k,l}, \theta_{i,k,l}^{s,v}}} \sum_{m=1}^{M} \sum_{k \in \mathcal{N}_i} \sum_{l \in \mathcal{N}_i \backslash \{k\}} U_{i,k,l}^m \left(e_{i,k,l} + e_{i,k} \right)$$

such that

$$C1 : U_{i,k,l}^m \in \{0, 1\}, \forall k, l \in \mathcal{N}_i$$
$$C2 : \sum_{k \in \mathcal{N}_i} U_{i,s,k}^m = \sum_{k \in \mathcal{N}_i} U_{i,k,s}^m = 1, \forall m \qquad (6.6)$$
$$C3 : \sum_{k,l \in \mathcal{N}_i} t_{i,k,l} U_{i,k,l}^m \le t_i^{\max}$$
$$C4 : t_{i,k} \le t_{i,k,l}, \sum_{k \in \mathcal{N}_i} f_{i,k} \le f_i^{\max}$$
$$C5 : \|v_{i,k,l}\|_2 \le u_i^{\max}, \|s_{i,k,l}\|_2 \le u_i^{\max}$$

where constraint C1 indicates that $U_{i,k,l}^m$ is a binary parameter; C2 indicates that the starting point of every route should be s, which is the position of the UAV; C3 ensures that the sum of the flight times of all the routes does not exceed $t_i^{\max}$; constraint C4 guarantees that the total computation time of the detection tasks for each turbine does not exceed the flight time, and the computation resources allocated for computing the detection tasks do not exceed the maximum computation resources $f_i^{\max}$ of the ith UAV; and C5 ensures that the airspeed and ground speed are each bounded by the maximum speed.

In problem (6.6), the optimal values of several parameters should be found. In addition, the variables at the upper bounds of the summation and the binary parameters make the problem difficult to be solved. To address the challenge, the solution of (6.6) is separated into two stages. In the first stage, the optimal UAV trajectory is found that minimizes the flight energy consumption by fixing the energy consumption $e_{i,k}$ of the UAV for task computation. We must calculate $s_{i,k,l}$, $v_{i,k,l}$, $t_{i,k,l}$, and $e_{i,k,l}$ for all $k, l \in \mathcal{N}_i$. The calculation of $s_{i,k,l}$ and $v_{i,k,l}$ differs, depending on whether the UAV is facing headwind or tailwind. The variable $\theta_{i,k,l}^{s,w}$ is utilized to determine wind conditions, and it can be calculated as the inner product of $s_{i,k,l}$ and w. The power consumption $p(v_{i,k,l})$ of a UAV flying with airspeed $v_{i,k,l}$ can be modeled according to [102]. With $s_{i,k,l}$, $v_{i,k,l}$, and $t_{i,k,l}$, the flight energy consumption $e_{i,k,l}$ of a UAV flying from turbine k to turbine l can be obtained by multiplying $t_{i,k,l}$ with $p(v_{i,k,l})$.

A heuristic algorithm is then designed to find the UAV's optimal trajectory. In particular, a brute force algorithm is developed to search for the optimal trajectory without considering time limits. The optimal trajectory obtained is then modified according to the maximum flight time. Based on the optimal trajectory found, a test is run to see if the UAV can fly back to its starting point when it decides to detect turbine l from turbine k. We now compare the cumulative flight time of a UAV from its starting point to turbine l via turbine k and the time it takes the UAV to fly from turbine l back to its starting point. If both of the times are less than the maximum time, $U_{i,k,l}^m = 1$; otherwise, $U_{i,k,l}^m = 0$, the UAV needs to fly back to the starting point from turbine k, and another detection round is added, starting from turbine l. In the second stage, based on the trajectory obtained, we solve the problem of the optimal computation resource allocation of the UAV detecting each turbine to minimize the energy consumption for computation. Since both the objective function and constraint C4 are convex, the Lagrange duality method can be utilized to solve this problem.

6.3.2.1 Performance Analysis

The results of the deployment are applied to show how to route the UAVs to inspect the wind turbines and execute the computation tasks. We use UAV 15 as an example. The UAV is placed at the turbine whose code is B110. Figure 6.5 shows the result of the trajectory. To minimize flight energy consumption, the UAV should avoid headwind. The UAV goes to C214 first and then chooses E105. After E105, the UAV uses the tailwind to go to A106 and A411. The total flight time for detection and computation is 15.3 min, and the energy consumption for the flight is 137.7 KJ. For comparison with the proposed method, a branch and bound method is used to find the optimal trajectory of the UAV according to (6.6). The total flight time calculated with the branch and bound method is 15.7 min, and the energy consumption for the flight is 141.5 KJ. Thus, the optimality of the proposed method can be proven.

Furthermore, Fig. 6.6 compares the computation energy consumption of the proposed method with that of the branch and bound method. Since constraint C4 in optimization problem (6.6) illustrates that the computation time of the detection tasks for each turbine must not exceed the flight time, and the flight time obtained by

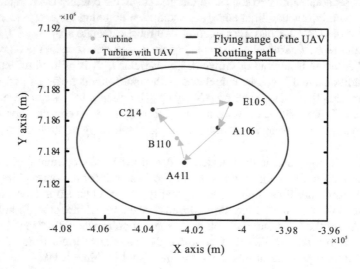

Fig. 6.5 Trajectory of UAV 15

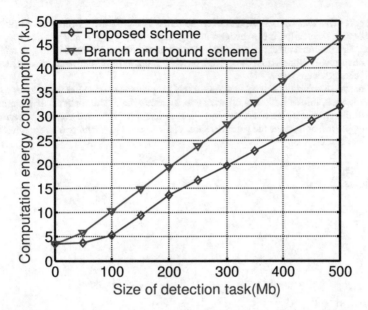

Fig. 6.6 Computation of the energy consumption of UAV 15 versus different sizes of detection tasks

the branch and bound method is longer than that obtained by the proposed method, we can conclude that the computation energy consumption of the proposed method is less than that of the branch and bound method. In addition, the computation energy consumption of the two methods increases with increasing detection task size.

6.4 Conclusions

In this chapter, we illustrated the application of UAV-assisted MEC in scenarios in which terrestrial MEC networks cannot be reliably established. According to the application scenarios, we analyzed the different roles of UAVs in MEC networks and presented the details for resource allocation and optimization in three UAV-assisted MEC network scenarios. In addition, we focused on UAV-assisted MEC at a wind farm and studied the wind's effect on UAVs' flight characteristics. The optimal number of UAVs to be deployed to inspect all the turbines at the wind farm was investigated. Then, the joint trajectory–computation resource optimization of the UAVs for inspection and task computation at the wind farm was studied. A two-stage method was developed to solve the problem of UAV trajectory optimization and computation resource allocation to detect each turbine, in order to minimize the total energy consumption.

Chapter 7
The Future of Mobile Edge Computing

Abstract This chapter first introduces the fundamental principles of blockchain and the integration of blockchain and mobile edge computing (MEC). Blockchain is a distributed ledger technology with a few desirable security characteristics. The integration of blockchain and MEC can improve the security of current MEC systems and provide greater performance benefits in terms of better decentralization, security, privacy, and service efficiency. Then, the convergence of artificial intelligence (AI) and MEC is presented. A federated learning–empowered MEC architecture is introduced. To improve the performance of the proposed scheme, asynchronous federated learning is proposed. The integration of blockchain and federated learning is also presented to enhance the security and privacy of the federated learning–empowered MEC scheme. Finally, more MEC enabled applications are discussed.

7.1 The Integration of Blockchain and Mobile Edge Computing (MEC)

MEC can offer a series of edge services with task processing, data storage, heterogeneity support, and QoS improvement capabilities. In close proximity to devices, MEC can provide instant computing applications with low latency and fast service response. The distributed structure of edge computing also potentially facilitates ubiquitous computing services, scalability, and network efficiency improvement. However, the MEC infrastructure still has unresolved challenges in terms of security and privacy. First, the large amount of heterogeneous data being collected, transferred, stored, and used in dynamic MEC networks can easily suffer serious data leakage. Further, due to the high dynamism and openness of MEC systems, it is very challenging to save the setting and configuration information of the edge servers in a reliable and secure way. Blockchain can enhance the security and privacy of MEC by offering many promising technical properties, such as decentralization, privacy, immutability, traceability, and transparency. The integration of blockchain and MEC can enable secure network orchestration, flexible resource management, and system performance improvements. In this section, we first introduce the structure of

© The Author(s) 2022
Y. Zhang, *Mobile Edge Computing*, Simula SpringerBriefs on Computing 9,
https://doi.org/10.1007/978-3-030-83944-4_7

blockchain and then present three potential cases of the integration of blockchain and MEC as future research directions.

7.1.1   The Blockchain Structure

Blockchain is an open database that maintains an immutably distributed ledger typically deployed in a peer-to-peer network. The structure of blockchain is shown in Fig. 7.1 and consists of three essential components: transactions, blocks of transaction records, and a consensus algorithm. The transaction information includes node pseudonyms, data types, metadata tags, a complete index history of metadata, an encrypted link to transaction records, and a timestamp of a specific transaction. Each transaction is encrypted and signed with digital signatures to guarantee authenticity. The digitally signed transactions are arbitrarily packed into a cryptographically tamper-evident data block. The blocks are linked in linear chronological order by hash pointers to form the blockchain. To maintain the consistency and order of the blockchain, a consensus algorithm is designed to generate agreement on the order of the blocks and to validate the correctness of the set of transactions constituting the block.

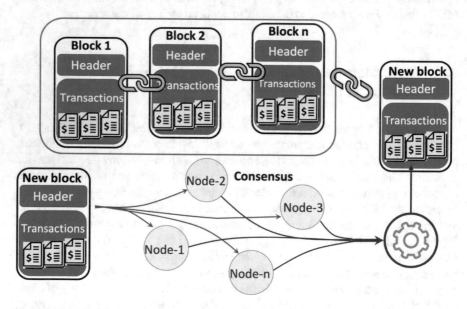

Fig. 7.1 The blockchain structure

7.1.1.1   Transactions

A transaction is the unit data structure of a blockchain, and it is created by a set of users to indicate the transfer of tokens from a specific sender to a receiver. Transactions generally consist of a recipient address, a sender address, and a token value. The input of a transaction is a reference that contains a cryptographic signature. The output of a transaction specifies an amount and an address. Transactions are bundled and broadcast to each node in the form of a block. As new transactions are distributed throughout the network, they are independently verified by each node. To protect the authenticity of a transaction, the functionalities of cryptographic hashing and asymmetric encryption are utilized, as follows.

- *Hash function*: A cryptographic hash function maps an arbitrary-size binary input into a fixed-size binary output. For example, SHA-256 maps an arbitrary-size input to a binary output 256 bits. The binary output is called a hash value. Moreover, the same input will always provide the same hash output. The probability of generating the same output for any two different inputs is negligible; it is thus impossible to reconstruct the input based on a hash output. The hash of a transaction makes it easy to keep track of transactions on the blockchain. In Bitcoin, SHA-256 and RIPEMD160 are utilized as hash function to produce a bitcoin address. In Ethereum, Keccak-256 is utilized as a hash function to produce a public key. In addition, signatures and private keys in blockchain frequently use hash functions to ensure security.
- *Asymmetric encryption*: Asymmetric encryption provides a secure method for authentication and confidentiality. Each node in a blockchain has a pair of keys: a public key and a private key. The public key can be shared with anyone to encrypt a message, whereas the private key should only be known to the key's initiator. In blockchain, the public key is used as the source address of transactions to verify their genuineness. The cryptographic private key is used to sign a transaction, which outputs a fixed-size digital signature for any arbitrary-size input message. The verification result will be true if the digital signature has the correct private key and input message. An elliptic curve digital signature algorithm is a typical algorithm for digital signing transactions. In the elliptic curve digital signature algorithm, when a user (A) wants to sign a transaction, that user first hashes the transaction and then uses his or her private key to encrypt the hashed transaction. The user then broadcasts the encrypted transaction. When another user receives the transaction and wants to verify its correctness, that user can decrypt the signature with user A's public key and hash the received transaction to verify whether the transaction information has been changed.

   In blockchain, each transaction is broadcast over the entire network and cross-verified by multiple nodes. The verified transactions are ordered consecutively with linearly ordered timestamps to guarantee correctness.

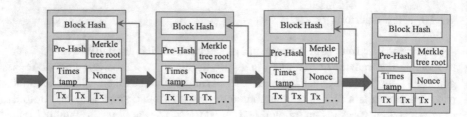

Fig. 7.2 Block structure

7.1.1.2   The Block Structure

A blockchain is a sequence of blocks that holds a complete list of transaction records. A block in a blockchain contains the hash of the current block, the hash of the previous block, a Merkle tree root, a timestamp, a nonce, and transactions, as shown in Fig. 7.2.

- *Block hash*: A block hash is the principal block identifier. It is a cryptographic digest made by hashing the block header twice with the SHA-256 algorithm. It identifies a block uniquely and unambiguously, and it can be independently derived by any node by simply hashing the block header.
- *Previous hash*: The hash of the previous block, which is a 256-bit hash that points to the previous block, is a necessary data field for the block header. Based on the previous block hash, all blocks are linked together to form a chain. If any block is tampered with, this will cause a change in all subsequent block hash pointers. When a block and all previous blocks are downloaded from an untrusted node, block hashing can be used to verify whether any block has been modified.
- *Merkle tree*: A Merkle tree represents a transaction set in the form of a binary tree for quick validation and synchronization. In the tree, the leaf nodes are the lowest tier of nodes, and each leaf node is a hash of a transaction. Each non-leaf node is a hash of the concatenation of two child nodes. The root node of the Merkle tree is known as the Merkle digest or root. Adjacent leaves are concatenated pairwise, and the hash of the concatenation constitutes the node's parent. Parent nodes are concatenated and hashed similarly to generate another level of parent nodes. This process is repeated until a single hash remains, which is the Merkle root. The Merkle tree is useful because it allows users to verify whether a transaction has occurred, based only on the direct branch from the transaction node to the Merkle root path. Moreover, the Merkle root allows tampering of any transaction data to be detected, to ensure their integrity.
- *Timestamp*: The time the block was generated. In blockchain, every block has a timestamp and the timestamp can be referred to as proof of existence. According to Satoshi Nakamoto's white paper [107], a decentralized timestamp service can resolve the double-spending problem. It can also help improve the traceability and transparency of the data stored in the blockchain.
- *Nonce*: A nonce is random number, and it can be used only once. Each node competes to find the nonce first to obtain the correct packing transactions for the

newly generated block. The nonce is difficult to find and is considered a way to weed out less talented miners. Once the nonce is found, it is added to the hashed block. With this number, the hash value of that block will be rehashed, creating a difficult algorithm.

The first block in any blockchain is termed the genesis block. This block is sometimes referred to as block 0. Every block in a blockchain stores a reference to the previous block. However, the genesis block has no previous block for reference.

7.1.1.3 Consensus Algorithms

Blockchain is a distributed decentralized network that provides immutability, privacy, security, and transparency. There is no central authority to validate and verify the transactions, but the transactions are considered secured and verified. This is possible because of consensus. Consensus is a process that allows all nodes to reach a common agreement on the state of a distributed ledger. The consensus problem can be formulated as a Byzantine fault–tolerant problem, that is, how generals can come to a common conclusion in the presence of a small number of traitors and miscommunications. The consensus currently used in most blockchain networks can be split into two categories: probabilistic-finality consensus and absolute-finality consensus. In probabilistic-finality consensus, any block in a blockchain can be reverted with a certain probability; attackers could thus accumulate a large amount of computational power, or stake, to create a long private chain to replace a valid chain. In absolute-finality consensus, a transaction is immediately finalized once it is included in a block. In other words, a new block generated by a leader node is committed by sufficient nodes before submission to the blockchain. We next present several common consensus strategies in blockchain.

- *Proof of work (PoW)*: PoW is a consensus strategy used in Bitcoin, where one node is selected to create a new block in each round of consensus through a computational power competition. In the competition, all participants must solve a cryptographic puzzle by using different nonces until the target is reached. The node that first solves the puzzle has the right to create a new block. Solving a PoW puzzle is costly and time-consuming, but it is easy for other nodes to verify. PoW guarantees security, based on the principle that it is impossible for a malicious attacker or group to collect more than 50% of the network's computational power to control the consensus process. PoW is a probabilistic-finality consensus protocol to guarantee eventual consistency. In PoW, nodes must consume a great deal of energy to solve the cryptographic puzzle. However, this work is useless and the energy consumed is wasteful. To tackle the resource waste problem of PoW, the idea of proof of useful resources was designed. Primecoin proposed a consensus algorithm to turn useless PoW into a meaningful search for special prime numbers when seeking a nonce [108]. Permacoin utilized bitcoin mining resources to distributively store an extremely large data provided by an authoritative file dealer based on proof of retrievability [109]. Instead of wasting energy for PoW, proof of burn allows miners to burn virtual currency tokens and then grants miners the right to write blocks in proportion to the number of burned coins [110].

- *Proof of stake (PoS)*: PoS is an energy-saving consensus to replace PoW. Instead of consuming large amounts of computational power to solve a PoW puzzle, PoS selects one node to create the next block based on the amount of stake. PoS is a probabilistic-finality consensus protocol, where the chances of being a block creator depends on "wealth". Since the richest node is bound to dominate the network, creator selection based on the amount of stake is quite unfair. Therefore, many researchers have proposed new schemes to decide on the node to forge the next block. Peercoin proposed a metric of coin age to measure the amount of held coins and their holding time [111]. In Peercoin, the node with older and larger sets of coins has a higher probability of creating the next block. Compared with PoW, Peercoin can reduce energy consumption and become more efficient. Ouroboros proposed PoS-based consensus, considering that stakes will shift over time [112]. A secure multiparty coin-flipping protocol was proposed in Ouroboros to guarantee the randomness of the leader election in the block generation process. To combine the benefits of PoW and PoS, proof of activity was proposed [113]. In proof of activity, the leader in each round of consensus is selected based on a standard PoW-based puzzle competition to generate an empty block header, where the stakeholders participating in the block verification receive a reward.
- *Delegated PoS (DPoS)*: The main difference between PoS and DPoS is that PoS involves direct democracy, whereas DPoS involves representative democracy [114]. In DPoS, stakeholders vote to elect delegates. The elected delegates are responsible for block creation and verification. Voting in DPoS is important, since it enables stakeholders to give delegates the right to create blocks, instead of creating blocks themselves; DPoS can thus reduce the computational power consumption of stakeholders to zero. On the other hand, PoW with plenty of nodes participating in the block verification process. In DPoS, only fewer delegates participate in the block verification process, thus the block can be confirmed quickly and the transactions can be confirmed quickly. Compared to PoW and PoS, DPoS is a low-cost, high-efficiency consensus protocol. Additionally, stakeholders do not need to worry about dishonest delegates, because these delegates can be easily voted out. There are also cryptocurrencies that implement DPoS, such as BitShares [115] and EoS. The new version of EoS has extended DPoS to DPoS–Byzantine fault tolerance. [116].
- Practical Byzantine fault tolerance (PBFT): PBFT is a Byzantine fault tolerance protocol with low algorithm complexity and high practicality [117]. Even if some nodes are faulty or malicious, network liveness and safety are guaranteed by PBFT, as long as a minimum percentage of nodes are connected, working properly, and behaving honestly. Hyperledger Fabric [118] utilizes PBFT as its consensus algorithm. In PBFT, a new block is determined in a round. In each round, a primary node is selected as the leader to broadcast the message sent by the client to other nodes. PBFT can be divided into three phases: pre-prepare, prepare, and commit. In each phase, a node enters the next phase if it has received votes from over two-thirds of all nodes. PBFT guarantees the nodes maintain a common state and take

Table 7.1 Main consensus comparison

	Type	Fault tolerance	Power consumption	Scalability
PoW	Probabilistic finality	50%	Large	Good
PoS	Probabilistic finality	50%	Less	Good
DPoS	Probabilistic finality	50%	Less	Good
PBFT	Absolute finality	33%	Negligible	Bad

consistent action in each round of consensus. PBFT achieves strong consistency and is thus an absolute-finality consensus protocol.

In distributed systems, there is no perfect consensus protocol. The consensus protocol should be adopted based on detailed application requirements. We present a simplified comparison of different consensus algorithms in Table 7.1.

7.1.2 Blockchain Classification

Current blockchain systems can be roughly classified into three types: public blockchains, consortium blockchains, and private blockchains. We compare these three types of blockchains from different perspectives.

- *Consensus determination*: In a public blockchain, each node can take part in the consensus process. In a consortium blockchain, only a selected set of nodes is responsible for validating a block. In a private blockchain, one organization fully controls and determines the final consensus.
- *Permission*: All transactions in a public blockchain are visible to the public. In a private or consortium blockchain, permissions depends on the organization or consortium decides whether the stored information is public or restricted.
- *Immutability*: Since transactions are stored in different nodes in the distributed network, it is nearly impossible to tamper with the public blockchain. However, if the majority of the consortium or the dominant organization wants to tamper with the blockchain, the consortium blockchain or private blockchain can be reversed or altered.
- *Efficiency*: It takes time to propagate transactions and blocks, since there are a large number of nodes in a public blockchain network. Taking network safety into consideration, restrictions on a public blockchain are much stricter. Therefore, transaction throughput is limited and latency is high. With fewer validators, consortium and private blockchains can be more efficient.

Table 7.2 Comparison of the different types of blockchains

	Public blockchain	Private blockchain	Consortium blockchain
Energy cost	High	Low	Low
Delay	Long	Short	Short
Security	High	Low	High

- *Centralization*: The main difference between the three types of blockchains is that a public blockchain is decentralized, a consortium blockchain is partially centralized, and a private blockchain is fully centralized, because it is controlled by a single group.
- *The consensus process*: Anyone can join the consensus process of a public blockchain. Different from public blockchains, both consortium and private blockchains are permissioned. A node needs to be certified to join the consensus process in consortium and private blockchains.

We compare the three types of blockchains in terms of energy costs, delay, and security, as shown in Table 7.2. Since a public blockchain often uses PoW to achieve consensus, it incurs high energy costs and long delays. A private blockchain is associated with low energy consumption and short delays to achieve consensus because of centralization. A consortium blockchain utilizes permissioned nodes to create new blocks without a mining process; it also therefore has low energy consumption and can achieve consensus quickly.

7.1.3   Integration of Blockchain and MEC

Many devices in MEC share their resources or content openly, without consideration of personal privacy. The integration of blockchain and MEC can establish a secure and private MEC system.

7.1.3.1   Blockchain for Edge Caching

With the rapid development of the Internet of Things (IoT) and wireless technologies, the huge amounts of data and content are undergoing exponential growth. To support massive content caching while also satisfying the low-latency requirements of content requesters, MEC provides distributed computing and caching resources in close proximity to users. Thus, content can be processed and then cached at the network edge, to alleviate data traffic on backhaul links and reduce content delivery latency. Since state-of-the-art devices are equipped with a certain amount of caching resources, a device with sufficient caching resources can be regarded as a caching

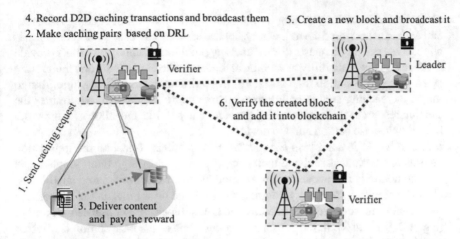

Fig. 7.3 Blockchain-empowered secure content caching

provider, to expand the caching capacity of the network edge. However, content usually involves the generator's sensitive personal information, such that devices might be not willing to store their content with an untrusted caching provider. A secure caching scheme among untrusted devices therefore needs to be built.

Blockchain enables untrusted nodes to interact with each other in a secure manner and provides a promising method for edge caching. We propose a blockchain-empowered distributed and secure content caching framework, as shown in Fig. 7.3 In this content caching system, devices can have two roles: a resource-constrained device with large-scale content is defined as a caching requester, and a device with sufficient caching resources is defined as a caching provider. Base stations are distributed in a specific area to work as edge servers. Specifically, if a content is successfully cached at one caching provider, the caching requester should create a transaction record and send it to the nearest base station. Base stations collect and manage local transaction records. The transaction records are structured into blocks after the consensus process among the base stations is completed and then stored permanently in each base station. The detailed processes are as follows.

- *System initialization*: For privacy protection, each device needs to register a legitimate identity in the system initialization stage. In an edge caching blockchain, an elliptic curve digital signature algorithm and asymmetric cryptography are used for system initialization. A device can obtain a legitimate identity after its identity has been authenticated. The identity includes a public key, a private key, and the corresponding certificate.
- *Roles in edge caching*: Devices choose their roles (i.e., caching requester and caching provider) according to their current caching resource availability state and future plans. Mobile devices with surplus caching resources can become caching providers to provide caching services for caching requesters.

- *Caching transactions*: Caching requesters send the amount of caching resources and expected serving time to the nearest base station. The base station broadcasts all received caching requests to local caching providers. Caching providers provide feedback on the availability of caching resources to the base station and their future plans. Each base station then utilizes a deep reinforcement learning algorithm to match the caching supply and demand pairs among the devices, determines the caching resources that each caching provider can provide, and allocate bandwidth between the base station and the devices.
- *Building blocks in a caching blockchain*: Base stations collect all the transaction records in a certain period and then encrypt and digitally sign them to guarantee their authenticity and accuracy. The transaction records are structured into blocks, and each block contains a cryptographic hash of the prior block in the consortium blockchain. To verify the correctness of a new block, the consensus algorithm (e.g., PBFT) is used. In the consensus process, one of the base station is selected as the leader for creating the new block. Because of broadcasts, each base station has access to the entire transaction record and has the opportunity to be the leader. In a consortium blockchain, the leader is chosen before the block building and does not change before the consensus process is completed.
- *The consensus process*: The leader broadcasts the created block to other base stations for verification and audit. All the base stations audit the correctness of the created block and broadcast their audit results. The leader then analyzes the audit results and, if necessary, sends the block back to the base stations for another audit. Following the audit results and corresponding signatures, compromised base stations will be discovered and held accountable.

The integration of blockchain and MEC can improve the security of edge networks and extend edge caching and resource sharing among untrusted entities.

7.1.3.2 Blockchain for Energy Trading

Due to harvesting and information communication technologies, distributed renewable energy sources are increasingly being integrated into smart grids, and vehicles not only can charge electricity from a home grid with renewable energy sources, but also can obtain electricity from other vehicles, to shift peak load through energy trading. However, because of privacy concerns, smart vehicles with surplus electricity might not be willing to work as energy suppliers in an energy trading market. To encourage vehicles with surplus electricity to participate in energy trading, the privacy of smart vehicles during the trade must be protected.

Blockchains, with its desirable characteristics of decentralization, immutability, accountability, and trustlessness, can significantly improve network security and save operational costs. Peer-to-peer topology enables electricity trading to be carried out in a decentralized, transparent, and secure market environment. The authors in [121] proposed a secure energy trading system with three types of components: vehicles, edge servers, and smart meters. The vehicles play three roles in electricity trading, with charging vehicles, discharging vehicles, and idle vehicles. Each vehicle chooses its own role based on its current energy state. Edge servers provide electricity and wireless communication services for the vehicles. Each charging vehicle sends a request about electricity demand to the nearest edge server. The edge server announces the energy demand to other vehicles (plug-in hybrid electric vehicles). Vehicles with surplus electricity submit selling prices to the edge server. After a double auction, two vehicles carry out an electricity trade. Smart meters are utilized to calculate and record the amount of electricity traded. Charging vehicles pay the discharging vehicles, based on the records in the smart meters. The detailed processes of the energy blockchain are similar to those in the caching blockchain, but there is still a very big difference. A caching blockchain utilizes a PBFT consensus algorithm, which requires relatively little energy and time to achieve consensus, because no mining process is involved. The work to achieve consensus is based on PoW. Although more energy and time must be spent for consensus, all the vehicles in a blockchain can participate in the process of verifying transactions, creating blocks, and achieving consensus.

7.2 Edge Intelligence: The Convergence of AI and MEC

The rapid development of AI techniques and applications has provided new possibilities for MEC. The integration of AI algorithms with MEC can considerably improve the intelligence and performance of edge computing. Conventional AI approaches rely on centralized mechanisms that invite serious security and privacy threats and are not suitable for resource-constrained edge networks. Federated learning and transfer learning are two emerging paradigms that shine new light on the convergence of AI and MEC.

7.2.1 Federated Learning in MEC

Increasing concerns of data privacy and security are hindering the wide implementation of AI algorithms to edge networks. Federated learning [122, 123] is proposed as a new learning scheme that enhances data privacy. Users participating in federated learning collaboratively train a global model and preserve their own data locally. Thus, by executing distributed training across users locally, federated learning enhances data privacy and reduces the cost of data transmission. By applying federated learning in MEC systems, the decision making process can be executed on edge devices, which reduces system latency and improves decision efficiency. Federated learning is believed to be one of the strongest enabling paradigms for large-scale MEC systems.

With the benefits of privacy enhancement, decentralization, and collaboration, federated learning has attracted significant attention in wireless networks. For exam-

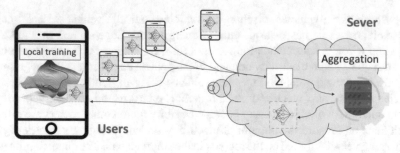

Fig. 7.4   Federated learning–empowered MEC

ple, Google exploited federated learning to train machine learning (ML) models for keyboard prediction [124]. Z. Yu et al. [125] proposed a federated learning–based proactive content caching scheme where the content caching policies are calculated by federated learning algorithms. However, in federated learning, the iterative communication between end users and the server and the local training of machine learning models by end users also consumes a large amount of resources.

To apply federated learning to MEC applications, a good volume of work has explored how to improve the performance of federated learning by optimizing the constrained resources in edge networks. J. H. Mills et al. [126] proposed adapting federated averaging [127] by adopting distributed Adam optimization to reduce the number of communication rounds for convergence. S. Wang et al. [128] proposed a control scheme to determine the optimal execution trade-off between local training and global aggregation within a given resource budget. In [129], C. Dinh et al. optimally allocated computation and communication resources in the network to improve the performance of federated learning deployed in wireless networks.

7.2.1.1   A Federated Learning–Empowered MEC Model

The architecture of federated learning–empowered MEC systems is depicted in Fig. 7.4. The end users in the system are the clients of federated learning, and the edge servers are the aggregation server of federated learning. For end user u_i with dataset D_i, the loss function for local training is defined as

$$F_i(w) = \frac{1}{|D_i|} \sum_{j \in D_i} f_j(w, x_j, y_j) \tag{7.1}$$

where $f_j(w, x_j, y_j)$ is the loss function on data sample (x_j, y_j) with parameter vector w, and $|D_i|$ is the size of the data samples in D_i. The loss function $f_j(w, x_j, y_j)$ is determined according to the specific learning algorithms, such as the mean squared error and the mean absolute error. The global loss function in federated learning is defined as

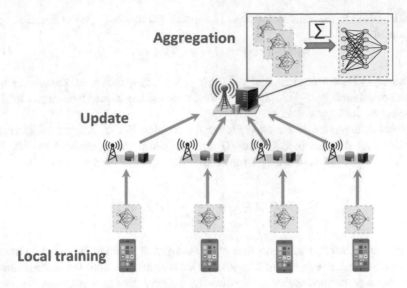

Fig. 7.5 Processes of federated learning–empowered MEC

$$F(w) = \frac{1}{|D|} \sum_{j \in D} f_j(h(w, x), y) = \frac{1}{|D|} \sum_i |D_i| \cdot F_i(w) \tag{7.2}$$

where $|D|$ is the size of the total training data $|D| = \sum_i |D_i|$. The objective of federated learning is to find the parameter vector w that minimizes the global loss function $F(w)$, that is,

$$Q(w, t) = \underset{i \in N, t \leq T}{\arg \min} F(w) \tag{7.3}$$

$$\text{such that} \quad \forall u_i \in U, i \in \{1, 2, \dots, N\} \tag{7.4}$$

where $u_i \in U$ denotes the user participating in the federated learning training process.

The general architecture of the federated learning–empowered MEC system consists of two planes: the end user plane and the edge server plane. As shown in Fig. 7.5, local training is executed in the user plane, while global aggregation is executed in the edge server plane. The federated learning–empowered MEC system involves three main steps: local training, parameter updating, and global aggregation. The MEC server plays the role of global server, and the end users, with mobile phones, smart vehicles, and IoT devices, and so on, are clients of federated learning. The three steps are repeated in the system to train the global machine learning model. Computation tasks are executed by running the federated learning algorithms in the MEC system.

- *Local training in the user plane:* The aggregation server distributes the ML model $\mathcal{M}$ to end users in the initialization phase. Each of the end users then trains the shared model $\mathcal{M}$ based on their local datasets. Gradient descent approaches are widely used in the training process. The model parameters $w_i(t)$ of iteration t are updated as

$$w_i(t) = w_i(t-1) - \eta \cdot \nabla F_i(w_i(t-1)), \tag{7.5}$$

where η is the learning rate, and $\nabla F_i(w_i(t-1))$ is the gradient of the loss function with parameters $w_i(t-1)$. The users then transmit the trained parameters $w(t)$ to the server for aggregation.

- *Global aggregation in the edge plane:* As denoted in Fig. 7.5, the MEC server collects all the parameters $\sum_i w_i(t)$ and calculates the aggregated model. The average aggregation is widely adopted to obtain the global model, as

$$w(t) = \frac{1}{\sum_{i=1}^{N} |D_i|} \sum_{i=1}^{N} |D_i| \cdot w_i(t) \tag{7.6}$$

The MEC server then transmits the aggregated global model to the end users to start a new training iteration. The learning process continues until the trained model reaches a predefined accuracy threshold or the execution time runs out.

7.2.1.2 Performance-Improved Asynchronous Federated Learning in MEC

Federated learning–empowered MEC systems can enlarge the scale of the training data and protect the data privacy of end users. However, new challenges have also arisen in the deployment of federated learning in MEC systems. First, the iterative update process of federated learning increases the transmission burden in communication resource–constrained edge networks. Second, the heterogeneous communication and computing capabilities of end users hinder the fast convergence of the learning process. Third, the risk of fake parameters from malicious participants also exists. To address these issues, a primary approach is to reduce the execution delay of federated learning. Thus, asynchronous federated learning is proposed.

In conventional federated learning, a synchronous mechanism is maintained by the clients and the global server to update the trained parameters and aggregate the global model. All the users participate in the global aggregation in each round. The training times of different end users varies greatly, because of their heterogeneous computing capabilities and dynamic communication states. In such a case, the execution time of each iteration is determined by the slowest clients, which incurs a high waiting cost for others, due to the heterogeneous runtimes. Asynchronous federated learning optimally selects a portion of the users to participate in global aggregation, while others continue with local training. Different optimizing approaches can be used as the node selection algorithm to decide on the participating nodes based on their capabilities. An overview of the asynchronous federated learning–empowered MEC scheme is shown in Fig. 7.6.

The asynchronous federated learning scheme comprises the following phases.

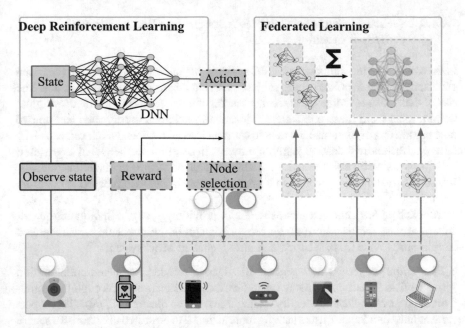

Fig. 7.6 Asynchronous federated learning–empowered MEC

- *Node selection:* Participating nodes are selected from all the end users through a node selection algorithm, according to their communication states and available computing resources. End users with sufficient resources are prone to being selected as participating nodes.
- *Local training and aggregation:* The participating nodes train their local models $m_i(t)$ according to their local data and obtain the parameters $w_i(t)$ for the trained model $m_i(t)$. User i also executes local aggregation by retrieving parameters $w_j(t)$ from nearby end users through device-to-device communication.
- *Global aggregation:* The MEC server carries out global aggregation based on local model parameters it has collected from participating end users, following Eq. (7.6). The global model $\mathcal{M}(t)$ is then broadcast to the end users to start a new learning iteration.

Deep reinforcement learning can be widely exploited as the node selection algorithm, deployed at the MEC server. The deep reinforcement learning algorithm learns the optimal node selection policy by using deep neural networks to approximate the policy gradient. Other techniques, such as convex optimization and game theory, can also be used in the node selection process.

7.2.1.3   Security-Enhanced AI in MEC: Integrating Blockchain with Federated Learning

In federated learning–empowered MEC systems, the parameters transmitted between end users and the MEC server are subject to serious security and privacy issues. The risk of data leakage increases, since an attacker can infer information on the original training data from these parameters. Moreover, malicious participants can upload fake parameters or use poisoned data to train their local models, which can cause the failure of the entire federated learning process. In addition, as the global aggregator, MEC servers also raise the risk of a single point of failure or malicious attacks. Building a trust mechanism among untrusted end users and MEC servers is therefore essential.

Blockchain has achieved great success in providing secure collaboration mechanisms among untrusted users. We propose integrating blockchain with federated learning to provide trust, security, and intelligence in MEC systems.

- *Blockchain for federated learning:* Blockchain provides a trusted collaboration mechanism for all participants (users) of federated learning. Through the authorization mechanism and identity management of the blockchain, especially a permissioned blockchain, users lacking mutual trust can be united to establish a secure and trusted cooperation mechanism. In addition, the model parameters of federal learning can be stored in the blockchain to ensure their safety and reliability.
- *Federated learning for blockchain:* The contradiction between the limited storage capacity of blockchain nodes and the larger storage demands of blockchains has always been a bottleneck in blockchain development. By processing the original data through federated learning, blockchains can store only the computation results, reducing storage cost and communication overhead. In addition, based on federated learning, the authentication calculation and transmission scheduling of blockchain transactions are optimized, which can considerably improve blockchain performance.

Based on the above analysis, we propose integrating blockchain with federated learning to build a trusted, secure, and intelligent MEC system. The integrated architecture is illustrated in Fig. 7.7. The architecture can be divided into the end user layer and the edge service layer. Users mainly consist of smart devices, such as IoT devices and mobile phones. The servers are represented by base stations equipped with MEC servers with certain storage and computing capabilities.

The integrated scheme consists of two main modules: federated learning and a permissioned blockchain. The federated learning learns the model parameters through local training on the user side, while the blockchain runs on the MEC server to collect and store the parameters of the federated learning. The parameters are verified by the consensus protocol. The detailed processes are as follows.

- *Local training:* Based on their local data, participating users train the model parameters through a gradient descent algorithm to minimize the loss function.

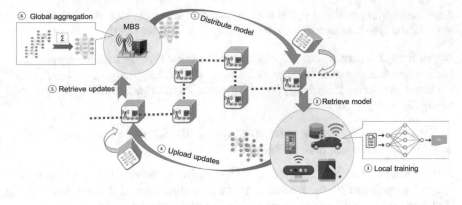

Fig. 7.7   The integration of blockchain and federated learning

- *Parameter transmission:* The trained local parameters are transmitted to the base station in the edge service layer through wireless links. The parameters of each user are collected and stored in blockchain nodes in the form of transactions.
- *Block generation:* Each blockchain node collects the transactions (model parameters) from the user layer and packages them into blocks using encryption and signatures. The block generator is determined by the consensus mechanism. The blockchain node that obtains the right to generate blocks broadcasts the block to the entire blockchain network and adds the block to the blockchain after verification.
- *Global aggregation:* The aggregator, that is, the MEC server, in the edge service layer aggregates model parameters according to the records in the blockchain and updates them into the global model. Furthermore, the global model is distributed to all participating users to start a new round of training.

The integration of blockchain and federated learning combines the security and trust of blockchains with the distributed intelligence of federated learning, which improves the security and data privacy of the MEC system.

7.2.2   Transfer Learning in MEC

7.2.2.1   Applying Transfer Learning in MEC

Transfer learning, as one of the machine learning methods, aims to transfer knowledge from existing domains to a new domain by learning across domains with non-independent and identically distributed data. Specifically, in transfer learning, a model developed for a task can be used as the original model for a related task. The basic idea of transfer learning is learning to learn, that is, to retain and reuse previously learned knowledge in the machine learning process. Different from traditional machine learning techniques, the source task and the target task are not the same, but related. The definition of transfer learning is as follows [130].

Definition 7.1 Given a source domain $\mathcal{D}_S$, a learning task $\mathcal{T}_S$, a target domain $\mathcal{D}_T$, and a target learning task $\mathcal{T}_T$, transfer learning aims to help improve the learning of the target predictive function $f_T(\cdot)$ in $\mathcal{D}_T$ using knowledge learned in $\mathcal{D}_S$ and $\mathcal{T}_S$, where $\mathcal{D}_S \neq \mathcal{D}_T$ or $\mathcal{T}_S \neq \mathcal{T}_T$.

To apply transfer learning to an MEC system, the following three main transfer learning research issues need to be addressed.

- *What to transfer:* Some knowledge can be specific to individual domains or tasks, while some knowledge can be common to both the source and target domains. It is therefore essential to determine which part of the knowledge can be transferred from the source domain to the target domain. The transferred knowledge helps to improve the performance of target tasks in the target domain.
- *How to transfer:* After determining what knowledge to transfer, learning algorithms or models need to be developed to transfer the knowledge from the source domain or source tasks to the target domain or target tasks.
- *When to transfer:* There are various applications and services in an MEC system. In some cases, the transfer of knowledge can improve system performance, while, in other cases, the transfer can decrease the quality of services or applications. Therefore, whether to transfer from the source domain to the target domain or not needs to be carefully analyzed.

In MEC systems, stochastic task models, heterogeneous MEC servers, and dynamic source data and user capabilities hinder the cooperation between the MEC servers, as well as the deployment of joint MEC tasks across different servers. To mitigate these challenges, transfer learning is believed to be a promising technique for deploying an MEC system across heterogeneous servers. Transfer learning–enabled MEC can be applied in the following scenarios.

- *Multiple computation tasks:* There can be multiple computation tasks and heterogeneous servers in MEC systems. Using transfer learning in the MEC systems can preserve the knowledge learned by some tasks and reuse it in related tasks. Servers in MEC systems can cooperate with each other by sharing and transferring the knowledge they learned from the local network within their coverage. Thus the utility of resources is improved and computation latency is reduced.
- *Computation offloading:* In computation offloading applications, optimal offloading strategies can be determined by learning a policy model with AI algorithms. By using transfer learning, the learned policy model can be used by other MEC servers as the starting point model. The model can be retrained for new MEC systems for a small computation cost. Thus the efficiency of computation offloading can be considerably improved, and energy consumptions can be further reduced.
- *Content caching:* In the content caching scenario, heterogeneous data types and dynamic caching capabilities among different MEC servers are the main issues in caching content across different MEC servers. The popularity of content in

different MEC systems can vary greatly due to the different types of users. Transfer learning can be exploited in heterogeneous MEC systems to mitigate the inaccuracy of caching models. The performance of the caching policy models and computation efficiency is thus improved in transfer learning–enabled MEC systems.

7.2.2.2   Federated Transfer Learning in MEC

Federated learning can connect isolated data and perform joint analyses on the data in way that preserves privacy. However, the issue of model personalization remains unresolved, since all users in federated learning share the same general model. In some cases, the general model might be not applicable to particular users. Moreover, the heterogeneous data distribution of users exacerbates the effective deployment of federated learning. To mitigate these issues, the concept of federated transfer learning emerges as a possible solution. The integration of federated learning and transfer learning broadens the application scope of federal learning. Applications with a small amounts of data or low-quality data can also obtain good machine learning models with the assistance of federated transfer learning.

Federated transfer learning differs from conventional transfer learning in the following aspects.

- *The training architectures are different.* Federated transfer learning is performed on distributed datasets from various users, and the original data are never transmitted to other users. Conventional transfer learning, however, can transmit the original data to a centralized server for training.
- *The machine learning models are trained in different places.* In federated transfer learning, the machine learning models are trained by all distributed users with their local computing resources and datasets. In conventional transfer learning, however, the training of ML models is usually completed by centralized servers.
- *The requirements for data security and privacy are different.* Federated transfer learning aims to protect the security and privacy of user data. In conventional transfer learning, the data face severe risks of leakage.

Research on federated transfer learning is still in its early stage. Y. Liu et al. [131] introduced a new framework, known as federated transfer learning, to improve the performance of machine learning models under a data federation. In [132], H. Yang et al. applied federated transfer learning to image steganalysis and proposed a framework named FedSteg to train a secure personalized distributed model on distributed user data. These limited works explored the integration of federated learning with transfer learning in various areas and provided rough frameworks of federated transfer learning. Federated transfer learning has huge potential in MEC in future networks. MEC can be enabled by federated transfer learning in the following areas.

- *Personalized services:* For future MEC systems, the provision of personalized services for different users is a crucial challenge. Enabled by federated transfer learning, knowledge of, for example, user behaviors and user preferences can be transferred among different users, based on the trained machine learning models. The quality of service in MEC systems can be considerably improved by the use of federated transfer learning techniques.

- *Super IoT:* The limited data storage capabilities and resources of IoT devices are major obstacles in deploying MEC systems in IoT networks. Federated transfer learning can mitigate the requirement for large amounts of data to train machine learning models. IoT devices can also train ML models with small amounts of data. Moreover, latency in training can be further reduced. The performance of IoT networks and applications can thus be improved.
- *Green communications:* With the increase in numbers of connected devices and applications, the energy costs of MEC systems are becoming a major concern that need to be addressed in future networks. Since machine learning models can be trained with small amounts of data, federated transfer learning decreases the computations for training and the communication overhead for data transmission. The energy cost is reduced and the efficiency is improved, leading to greener communications in future networks.

7.3   MEC in Other Applications

Along with the advancement of beyond 5G technology and the pervasive IoT, MEC techniques have been applied in diverse scenarios to meet intensive computing, processing, and analysis demands for disease prevention, industrial production, and emergency response.

7.3.1   MEC in Pandemics

A pandemic is the spread of an infectious disease across large regions, and it poses serious health risks to huge numbers of people. For instance, the recent COVID-19 pandemic has infected more than 20 million people worldwide and has severely affected the global economy. Early estimates demonstrated that most major economies lost at least 2.4% of their gross domestic product during 2020.

Since pandemic outbreaks are always a surprise and people are largely unprepared to address them, in the early stage, a virus has an extraordinary capacity to spread. It is therefore imperative to establish a pandemic prediction system that can provide valuable and comprehensive information for pre-judging the time, location, scale, and other key characteristics of virus outbreaks.

The efficient operation and precise judgment of the pandemic prediction system are based on comprehensive data collection and complex data analysis. The data are captured by sensing devices that are widely placed in crowded areas, such as bus stations, shopping malls, and schools, and are characterized by large sizes, diverse types, and heterogeneous elements. Processing the captured data with feature extrac-

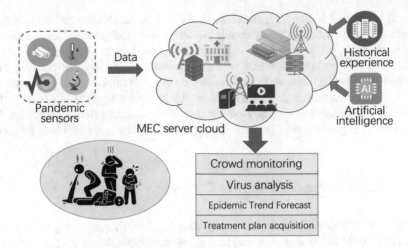

Fig. 7.8 MEC in pandemics

tion and correlation analysis always requires large amounts of computing resources. A traditional approach to meet this demand is to upload the data to remote cloud servers for processing. However, a great deal of communication costs involve yields in this transmission, especially in wireless remote access scenarios. Moreover, the cloud-based processing approach can incur long time delays, which fails to cater to the fast response requirements of applications such as personnel monitoring at a station entrance.

MEC is an appealing paradigm to address this problem. It helps pandemic prediction systems process sensing data in proximity to virus monitoring areas and provides important epidemiological reports about virus spread trends in a short time. MEC servers are equipped on cellular network base stations, Wi-Fi access points, and other nodes that facilitate data transmission and have a stable energy supply. With the continuous enhancement of AI technology, which is capable of revealing hidden issues and correlations from big data, the incorporation of AI and MEC has emerged as a promising approach. Thus, machine learning modules should be deployed to MEC servers to track diseases, predict their growth, and propose coping strategies. For instance, an MEC-empowered deep learning model is suitable for disease classification, while MEC-aided multi-agent learning can be used to predict infection trends in large areas. Moreover, since pandemic prediction and prevention require joint analyses and actions between different departments in multiple regions, a multi-level collaborative MEC service system must be created that shares virus information among MEC servers to better understand and address the pandemic crisis. Figure 7.8 shows a typical framework of MEC techniques applied to pandemic prediction.

In applying MEC in pandemics, open questions still exist. The first involves the privacy protection of the MEC service. In pandemic prediction, the monitoring target is human activities and their physical characteristics, through which malicious users could obtain individuals' private information, such as personal daily activity trajectories and health status. Recent research has indicated that nontraditional data sources, including social media, Internet search histories, and smartphone data, which are closely related to privacy, are helpful in forecasting pandemic. Consequently, MEC-empowered pandemic management with strict privacy protection is imperative. Furthermore, for the flexible and dynamic detection of pandemics in multiple locations, the pandemic monitoring devices should be lightweight and portable. If the MEC service is integrated into a battery-powered mobile pandemic detection device, the energy efficiency of the data processing will become a key issue for consideration.

7.3.2 MEC in the Industrial IoT (IIoT)

Enabled by IoT technology, industry has witnessed substantial changes in operational efficiency, product quality, and management mechanisms in recent years, and it is continuously evolving toward the IIoT. From the perspective of manufacturers, the proliferation of the IIoT will provide interconnections between large-scale distributed industrial equipment, enable a comprehensive awareness of production environments, and help realize full industrial automation.

Along with the evolution of the IIoT, large amounts of data regarding factory environment status detection, robot device control, and product quality monitoring are being generated and processed. Since modern industrial production is an assembly line operation, any instruction error or behavior lag in the production process will seriously affect the overall manufacturing efficiency. Consequently, the demand for data processing services of high reliability and low latency has increased.

MEC technology, which can facilitate data processing closer to industrial facilities, thereby enabling production managers and equipment controllers to speed up their decision making, has been widely recognized as a promising approach to cater to the demands mentioned. Figure 7.9 shows typical scenarios of MEC application in industrial automation control, logistics transportation management, product quality assurance, and energy scheduling.

To boost production efficiency, remain profitable, and replace expensive human labor with ever-cheaper machines, various manufacturing robots are being widely used in industrial factories. During the operation of the robots, MEC servers work as information processors and control systems that analyze the robot monitoring data from sensors and actuators, while generating control instructions for robotic arm behavior and coping with problems in automated production lines.

Smart logistics have become a key attribute of modern industry. They incorporate autonomous transport vehicles, sensor-driven cargo tracking tools, and online automated sales platforms throughout the whole supply and sale chain. With the aid of MEC technology, unmanned vehicles can achieve more precise and real-time driving control, the transportation status of cargo can be tracked throughout the process, and sales strategies can be optimized in time.

Product quality is the core element of industrial production, and there are many quality inspection methods. With the development of AI, machine learning

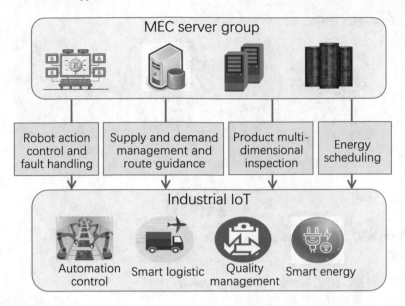

Fig. 7.9 MEC in the IIoT

approaches have been introduced to identify the characteristics of products' dimensions, performance, and stability. The learning process always requires intense data processing and complex model construction. MEC servers that provide sufficient computing capabilities at the site of quality inspection facilities are crucial elements to cater to this requirement.

Industrial manufacturing relies heavily on energy consumption. Among the diversified energy types, electrical energy has been proven to have the most important effects on factory production capacities and costs. With the rise of smart grids, the matching of electricity supply and demand has become flexible, but has also created calculation demands, such as for grid state analysis and user demand trend prediction. MEC is an appealing approach to address this additional demand. Furthermore, besides traditional energy types that harm the environment, renewable energy sources, such as solar, wind, and tidal energy, are beginning to be widely used in industrial production. The time-varying and unstable supply characteristics of renewable energy also require MEC's analytical monitoring and adaptive scheduling. Although MEC technology provides many benefits to the IIoT, some challenges of industrial MEC remain unresolved. For instance, the MEC-empowered IIoT is vulnerable to malicious attacks. Since wireless has been pervasively used in IIoT device-to-device communication, task offloading data can be easily eavesdropped and forged, resulting in the leakage of commercial secrets or production interruptions. In addition, industrial logistic vehicles move throughout large geographical areas and can therefore access heterogeneous MEC servers. The coordination and integration of MEC services is also an unexplored issue.

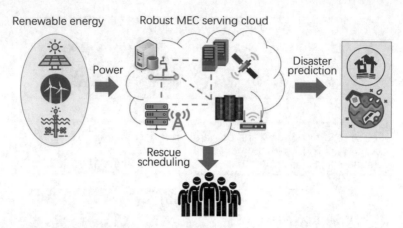

Fig. 7.10 MEC in disaster management

7.3.3 MEC in Disaster Management

Sudden disasters cause serious and widespread human, economic, or environmental losses. To address this issue, disaster management has been proposed for taking some countermeasures and scheduling relief supplies to protect human lives and infrastructures.

To ensure the effective operation of a disaster management system, a large amount of information needs to be processed, which is mainly reflected in two aspects. The first involves the comprehensive analysis of collected environmental data, including meteorological, geological, and hydrological data, to accurately predict possible disasters. On the other hand, after the occurrence of a disaster, progress monitoring and estimations of the status of the disaster relief and supply of materials are required to facilitate the scheduling of rescuers and resources. To meet these information processing demands, servers with powerful computing capabilities should be equipped in the disaster areas. Due to possible damage to communication network facilities and lines caused by the disaster, core cloud servers and remote task offloading are not suitable for providing computing services. MEC's proximity computing service can effectively make up for these shortcomings. However, a single MEC server can also be damaged in a severe disaster; therefore, a group of distributed MEC servers empowered with robustness and survivability is a feasible solution.

Figure 7.10 illustrates the framework of an MEC-empowered disaster management system, where each MEC server pair is connected through several redundant backup communication lines. These links can be wired connections or wireless connections through cellular networks, Wi-Fi, or even satellite networks. The dual backup capability of the servers and communication links greatly improves the robustness and survivability of the entire MEC system. To cope with a potentially unstable power supply in the disaster area, MEC servers can leverage renewable energy and use energy batteries as storage devices to adapt to the time-varying characteristics of wind and solar power. In addition, MEC servers are evolving toward miniaturization and lightweight configurations to meet the portability requirements of a disaster relief operation carried out at multiple locations.

Despite the advantages that edge computing has provided disaster management, key issues remain unexplored in MEC service deployment. A typical problem involves the energy efficiency of MEC servers. Due to the lack of energy supply in disaster areas and the constrained battery power of portable servers, providing powerful computing capabilities at a low energy cost is a critical challenge. Moreover, the effective integration of diversified disaster environment detection networks and heterogeneous rescue systems with MEC services urgently requires further investigation.

References

1. C. Perera, A. Zaslavsky, P. Christen, D. Georgakopoulos, Context aware computing for the internet of things: a survey. IEEE Commun. Surv. Tutor. **16**(1), 414–454 (2014)
2. IBM News Releases, IBM and Nokia Siemens Networks announce world first mobile edge computing platform. (2013). http://www-03.ibm.com/press/us/en/pressrelease/40490.wss
3. Mobile-edge computing—Introductory technical white paper, White Paper, ETSI, Sophia Antipolis, France, September (2014). https://portal.etsi.org/portals/0/tbpages/mec/docs/mobile-edge_computing_-_introductory_technical_white_paper_v1%2018-09-14.pdf
4. M. Chen, W. Saad, C. Yin, Virtual reality over wireless networks: quality-of-service model and learning-based resource management. IEEE Trans. Commun. **66**(11), 5621–5635 (2018). https://doi.org/10.1109/TCOMM.2018.2850303
5. Y. Sun, Z. Chen, M. Tao, H. Liu, Communications, caching, and computing for mobile virtual reality: modeling and tradeoff. IEEE Trans. Commun. **67**(11), 7573–7586 (2019)
6. A. Al-Shuwaili, O. Simeone, Energy-efficient resource allocation for mobile edge computing-based augmented reality applications. IEEE Wirel. Commun. Lett. **6**(3), 398–401 (2017)
7. Y. Dai, D. Xu, S. Maharjan, Y. Zhang, Joint load balancing and offloading in vehicular edge computing and networks. IEEE Int. Things J. **6**(3), 4377–4387 (2019). https://doi.org/10.1109/JIOT.2018.2876298
8. N. Abbas, Y. Zhang, A. Taherkordi, T. Skeie, Mobile edge computing: a survey. IEEE Int. Things J. **5**(1), 450–465 (2018). https://doi.org/10.1109/JIOT.2017.2750180
9. F.Y. Okay, S. Ozdemir, A fog computing based smart grid model, in *International Symposium on Networks. Computers and Communications*, Yasmine, Hammamet, (2016), pp. 1–6. https://doi.org/10.1109/ISNCC.2016.7746062
10. Y. Mao, C. You, J. Zhang, K. Huang, K.B. Letaief, A survey on mobile edge computing: the communication perspective. IEEE Commun. Surv. Tutor. **19**(4), 2322–2358 (2017). https://doi.org/10.1109/COMST.2017.2745201
11. J. Qiao, Y. He, X.S. Shen, Proactive caching for mobile video streaming in millimeter wave 5G networks. IEEE Trans. Wirel. Commun. **15**(10), 7187–7198 (2016)
12. S. Zhou, Y. Sun, Z. Jiang, Z. Niu, Exploiting moving intelligence: delay-optimized computation offloading in vehicular fog networks. IEEE Commun. Mag. **57**(5), 49–55 (2019). https://doi.org/10.1109/MCOM.2019.1800230
13. P. Mach, Z. Becvar, Mobile edge computing:a survey on architecture and computation offloading, in *IEEE Communications Surveys & Tutorials*, vol. 19, no. 3, (2017), pp. 1628–1656
14. Y. Dai, D. Xu, S. Maharjan, Y. Zhang, Joint computation offloading and user association in multi-task mobile edge computing. IEEE Trans. Veh. Technol. **67**(12), 12313–12325 (2018)

Y. Zhang, *Mobile Edge Computing*, Simula SpringerBriefs on Computing 9,
https://doi.org/10.1007/978-3-030-83944-4

15. J. Zhang, X. Hu, Z. Ning et al., Energy-latency tradeoff for energy-aware offloading in mobile edge computing networks. IEEE Int. Things J. **5**(4), 2633–2645 (2018). https://doi.org/10.1109/JIOT.2017.2786343

16. M. Deng, H. Tian, B. Fan, Fine-granularity based application offloading policy in cloud-enhanced small cell networks, in *IEEE International Conference on Communications*, Kuala Lumpur, Malaysia, 23–27 May 2016 (IEEE, 2016), pp. 638–643

17. W. Zhang, Y. Wen, D.O. Wu, Collaborative task execution in mobile cloud computing under a stochastic wireless channel. IEEE Trans. Wirel. Commun. **14**(1), 81–93 (2015)

18. J. Liu, Y. Mao, J. Zhang, K.B. Letaief, Delay-optimal computation task scheduling for mobile-edge computing systems, in *IEEE International Symposium on Information Theory*, Barcelona, (2016), pp. 1451–1455

19. J. O. Fajardo, I. Taboada, F. Liberal, Radio-aware service-level scheduling to minimize downlink traffic delay through mobile edge computing, in *International Conference on Mobile Networks and Management*, Santander, 16–18 Sep 2015 (Springer, 2015), pp. 121–134

20. Y. Mao, J. Zhang, K.B. Letaief, Dynamic computation offloading for mobile-edge computing with energy harvesting devices. IEEE J. Sel. Areas Commun. **34**(12), 3590–3605 (2016)

21. M. Kamoun, W. Labidi, M. Sarkiss, Joint resource allocation and offloading strategies in cloud enabled cellular networks, in *IEEE International Conference on Communications*, London, 8–12 June 2015 (IEEE, 2015), pp. 5529–5534

22. W. Labidi, M. Sarkiss, M. Kamoun, Energy-optimal resource scheduling and computation offloading in small cell networks, in *IEEE 22nd International Conference on Telecommunications*, Sydney, 27–29 Apr 2015 (IEEE, 2015), pp. 313–318

23. W. Labidi, M. Sarkiss, M. Kamoun, Joint multi-user resource scheduling and computation offloading in small cell networks, in *IEEE International Conference on Wireless and Mobile Computing, Network, and Communications*, Abu Dhabi, 19–21 Oct 2015 (IEEE, 2015), pp. 794–801

24. K. Zhang, Y. Mao, S. Leng, Q. Zhao, L. Li, X. Peng, L. Pan, S. Maharjan, Y. Zhang, Energy-efficient offloading for mobile edge computing in 5G heterogeneous networks. IEEE Access **4**, 5896–5907 (2016)

25. M. Chen, B. Liang, M. Dong. Joint offloading and resource allocation for computation and communication in mobile cloud with computing access point, in *2017 IEEE International Conference on Computer Communications*, Atlanta, Georgia, 1–4 May 2017 (IEEE, 2017), pp. 1–6

26. T.Q. Dinh, J. Tang, Q.D. La, T.Q.S. Quek, Offloading in mobile edge computing: task allocation and computational frequency scaling. IEEE Trans. Commun. **65**(8), 3571–3584 (2017)

27. Y. Wang, M. Sheng, X. Wang, L. Wang, J. Li, Mobile-edge computing: partial computation offloading using dynamic voltage scaling. IEEE Trans. Commun. **64**(10), 4268–4282 (2016)

28. J. Ren, G. Yu, Y. Cai, Y. He, F. Qu, Partial offloading for latency minimization in mobile-edge computing, in *2017 IEEE Global Communications Conference*, Singapore, 4–8 Dec 2017 (IEEE, 2017), pp. 1–6

29. U. Saleem, Y. Liu, S. Jangsher, and Y. Li, Performance guaranteed partial offloading for mobile edge computing, in *2018 IEEE Global Communications Conference*, Abu Dhabi, 9–13 Dec 2018 (IEEE, 2018), pp. 1–6

30. S.E. Mahmoodi, R.N. Uma, K.P. Subbalakshmi, Optimal joint scheduling and cloud offloading for mobile applications. IEEE Trans. Cloud Comput. **7**(2), 301–313 (2019). https://doi.org/10.1109/TCC.2016.2560808

31. W. Zhang, Y. Wen, D.O. Wu, Collaborative task execution in mobile cloud computing under a stochastic wireless channel. IEEE Trans. Wirel. Commun. **14**(1), 81–93 (2015)

32. O. Muñoz, A. Pascual-Iserte, J. Vidal, *Joint Allocation of Radio and Computational Resources in Wireless Application Offloading* (Future Network & Mobile Summit, Lisbon, 2013), pp. 1–10

33. O. Muñoz, A. Pascual-Iserte, J. Vidal, Optimization of radio and computational resources for energy efficiency in latency-constrained application offloading. IEEE Trans. Veh. Technol. **64**(10), 4738–4755 (2015)

34. Y. Mao, J. Zhang, S.H. Song, K.B. Letaief, Power-delay trade-off in multi-user mobile-edge computing systems, in *IEEE Global Communications Conference*, Washington, DC, 4–8 Dec 2016 (IEEE, 2016), pp. 1–6

35. X. Xu, M. Tao, C. Shen, Collaborative multi-agent multi-armed bandit learning for small-cell caching. IEEE Trans. Wirel. Commun. **19**(4), 2570–2585 (2020)

36. F. Wang, F. Wang, J. Liu, R. Shea, L. Sun, Intelligent video caching at network edge: a multi-agent deep reinforcement learning approach, in *IEEE International Conference on Computer Communications* (2020). [AB25]

37. G. Qiao, S. Leng, S. Maharjan, Y. Zhang, N. Ansari, Deep reinforcement learning for cooperative content caching in vehicular edge computing and networks. IEEE Int. Things J. **7**(1), 247–257 (2020)

38. R. Karasik, O. Simeone, S. Shamai, How much can D2D communication reduce content delivery latency in fog networks with edge caching? IEEE Trans. Commun. **68**(4), 2308–2323 (2020)

39. K. Zhang, J. Cao, H. Liu, S. Maharjan, Y. Zhang, Deep reinforcement learning for social-aware edge computing and caching in urban informatics. IEEE Trans. Ind. Inf. **16**(8), 5467–5477 (2020)

40. W. Wu, N. Zhang, N. Cheng, Y. Tang, K. Aldubaikhy, X. Shen, Beef up MM Wave dense cellular networks with D2D-assisted cooperative edge caching. IEEE Trans. Veh. Technol. **68**(4), 3890–3904 (2019)

41. N. Zhao, X. Liu, Y. Chen, S. Zhang, Z. Li, B. Chen, M. Alouini, Caching D2D connections in small-cell networks. IEEE Trans. Veh. Technol. **67**(12), 12326–12338 (2018)

42. R. Zhang, F.R. Yu, J. Liu, T. Huang, Y. Liu, Deep reinforcement learning (DRL)-based device-to-device (D2D) caching with blockchain and mobile edge computing. IEEE Trans. Wirel. Commun. **19**(10), 6469–6485 (2021)

43. Y. Saputra, D. Hoang, D. Nguyen, E. Dutkiewicz, D. Niyato, D. Kim, Distributed deep learning at the edge: a novel proactive and cooperative caching framework for mobile edge networks. IEEE Wirel. Commun. Lett. **8**(4), 1220–1223 (2019)

44. J. Kwak, Y. Kim, L. Le, S. Chong, Hybrid content caching in 5G wireless networks: cloud versus edge caching. IEEE Trans. Wirel. Commun. **17**(5), 3030–3045 (2018)

45. S. Dang, O. Amin, B. Shihada, M. Alouini, What should 6G be? Nat. Electron. **3**(1), 20–29 (2020.) https://doi.org/10.1038/s41928-019-0355-6

46. W. Saad, M. Bennis, M. Chen, A vision of 6G wireless systems: applications, trends, technologies, and open research problems. IEEE Netw. **34**(3), 134–142 (2020). https://doi.org/10.1109/MNET.001.1900287

47. K.B. Letaief, W. Chen, Y. Shi, J. Zhang, Y.A. Zhang, The roadmap to 6G: AI empowered wireless networks. IEEE Commun. Mag. **57**(8), 84–90 (2019). https://doi.org/10.1109/MCOM.2019.1900271

48. K. Zhang, Y. Zhu, S. Maharjan, Y. Zhang, Edge intelligence and blockchain empowered 5G beyond for the Industrial Internet of Things. IEEE Netw. **33**(5), 12–19 (2019). https://doi.org/10.1109/MNET.001.1800526

49. Y. Lu, X. Huang, K. Zhang, S. Maharjan, Y. Zhang, Blockchain and federated learning for 5G beyond. IEEE Netw. **35**(1), 219–225 (2021). https://doi.org/10.1109/MNET.011.1900598

50. P. Mach, Z. Becvar, Mobile edge computing: a survey on architecture and computation offloading. IEEE Commun. Surv. Tutor. **19**(3), 1628–1656 (2017). https://doi.org/10.1109/COMST.2017.2682318

51. Multi-access edge computing (MEC); Phase 2: use cases and requirements, Standard ETSI GS MEC 002 V2.1.1 (ETSI MEC Group, 2018)

52. Y. Lu, X. Huang, K. Zhang, S. Maharjan, Y. Zhang, Blockchain empowered asynchronous federated learning for secure data sharing in Internet of Vehicles. IEEE Trans. Veh. Technol. **69**(4), 4298–4311 (2020). https://doi.org/10.1109/TVT.2020.2973651

53. S. Zhou, Y. Sun, Z. Jiang, Z. Niu, Exploiting moving intelligence: delay-optimized computation offloading in vehicular fog networks. IEEE Commun. Mag. **57**(5), 49–55 (2019)

54. L. Pu, X. Chen, G. Mao, Q. Xie, J. Xu, Chimera, an energy-efficient and deadline-aware hybrid edge computing framework for vehicular crowdsensing applications. IEEE Int. Things J. **6**(1), 84–99 (2019)
55. Z. Zhou, J. Feng, Z. Chang, X. Shen, Energy-efficient edge computing service provisioning for vehicular networks: a consensus ADMM approach. IEEE Trans. Veh. Technol. **68**(5), 5087–5099 (2019)
56. X. Li, Y. Dand, M. Aazam, X. Peng, T. Chen, C. Chen, Energy-efficient computation offloading in vehicular edge cloud computing. IEEE Access **8**, 37632–37644 (2020)
57. J. Ren, G. Yu, Y. He, G.Y. Li, Collaborative cloud and edge computing for latency minimization. IEEE Trans. Veh. Technol. **68**(5), 5031–5044 (2019)
58. C. Lin, G. Han, X. Qi, M. Guizani, L. Shu, A distributed mobile fog computing scheme for mobile delay-sensitive applications in SDN-enabled vehicular networks. IEEE Trans. Veh. Technol. **69**(5), 5481–5493 (2020)
59. Y. Ku, P. Chiang, S. Dey, Real-time QoS optimization for vehicular edge computing with off-grid roadside units. IEEE Trans. Veh. Technol. **69**(19), 11975–11991 (2021)
60. X. Hou, Z. Ren, J. Wang, W. Cheng, Y. Ren, K. Chen, H. Zhang, Reliable computation offloading for edge-computing-enabled software-defined IoV. IEEE Int. Things J. **7**(8), 7097–7111 (2020)
61. X. Sun, J. Zhao, X. Ma, Q. Li, Enhancing the user experience in vehicular edge computing networks: an adaptive resource allocation approach. IEEE Access **7**, 161074–161087 (2019)
62. L. Zhao, K. Yang, Z. Tan, X. Li, S. Sharma, Z. Liu, A novel cost optimization strategy for SDN-enabled UAV-assisted vehicular computation offloading, in *IEEE Transactions on Intelligent Transportation Systems*, forthcoming (2021)
63. J. Du, F.R. Yu, X. Chu, J. Feng, G. Lu, Computation offloading and resource allocation in vehicular networks based on dual-side cost minimization. IEEE Trans. Veh. Technol. **68**(2), 1079–1092 (2019)
64. Z. Deng, Z. Cai, M. Liang, A multi-hop VANETs-assisted offloading strategy in vehicular mobile edge computing. IEEE Access **8**, 53062–53071 (2020)
65. Y. Hui, Z. Su, T.H. Luan, C. Li, Reservation service: trusted relay selection for edge computing services in vehicular networks. IEEE J. Sel. Areas Commun. **38**(12), 2734–2746 (2021)
66. H. Liu, P. Zhang, G. Pu, T. Yang, S. Maharjan, Y. Zhang, Blockchain empowered cooperative authentication with data traceability in vehicular edge computing. IEEE Trans. Veh. Technol. **69**(4), 4221–4232 (2020)
67. J. Zhang, H. Zhong, J. Cui, M. Tian, Y. Xu, L. Liu, Edge computing-based privacy-preserving authentication framework and protocol for 5G-enabled vehicular networks. IEEE Trans. Veh. Technol. **69**(7), 7940–7954 (2020)
68. Y. Liu, S. Wang, Q. Zhao, S. Du, A. Zhou, X. Ma, F. Yang, Dependency-aware task scheduling in vehicular edge computing. IEEE Int. Things J. **7**(6), 4961–4971 (2020)
69. J. Zhang, H. Guo, Y. Zhang, Task offloading in vehicular edge computing networks: a load-balancing solution. IEEE Trans. Veh. Technol. **69**(2), 2092–2104 (2020)
70. C. Yang, Y. Liu, X. Chen, W. Zhong, S. Xie, Efficient mobility-aware task offloading for vehicular edge computing networks. IEEE Access **7**, 26652–26664 (2019)
71. S. Buda, S. Culeng, C. Wu, J. Zhang, K.A. Yau, Y. Ji, Collaborative vehicular edge computing towards greener ITS. IEEE Access **8**, 63935–63944 (2020)
72. S.S. Shan, M. Ali, A.W. Malik, M.A. Khan, S.D. Ravana, vFog: a vehicle-assisted computing framework for delay-sensitive applications in smart cities. IEEE Access **7**, 34900–34909 (2019)
73. Y. Hui, Z. Su, T.H. Luan, J. Cai, Content in motion: an edge computing based relay scheme for content dissemination in urban vehicular networks. IEEE Trans. Intell. Transp. Syst. **20**(8), 3115–3128 (2019)
74. Q. Qi, J. Wang, Z. Ma, Y. Cao, L. Zhang, J. Liao, Knowledge-driven service offloading decision for vehicular edge computing: a deep reinforcement learning approach. IEEE Trans. Veh. Technol. **68**(5), 4192–4203 (2019)

75. C. Sonmez, C. Tunca, A. Ozgovde, C. Ersoy, Machine learning-based workload orchestrator for vehicular edge computing, in *IEEE Transactions on Intelligent Transportation Systems*, forthcoming (2020)

76. D. Chen, Y. Liu, B. Kim, J. Xie, C. Hong, Z. Han, Edge computing resources reservation in vehicular networks: a meta-learning approach. IEEE Trans. Veh. Technol. **69**(5), 5634–5646 (2020)

77. W. Zhan, C. Luo, J. Wang, C. Wang, G. Min, H. Duan, Q. Zhu, Deep-reinforcement-learning-based offloading scheduling for vehicular edge computing. IEEE Int. Things J. **7**(6), 5449–5465 (2020)

78. X. Wang, Z. Ning, S. Guo, L. Wang, Imitation learning enabled task scheduling for online vehicular edge computing, in *IEEE Transactions on Mobile Computing*, forthcoming (2021)

79. K. Zhang, Y. Mao, S. Leng, Y. He, Y. Zhang, Mobile-edge computing for vehicular networks: a promising network paradigm with predictive off-loading. IEEE Veh. Technol. Mag. **12**(2), 36–44 (2017)

80. K. Zhang, Y. Zhu, S. Leng, Y. He, S. Maharjan, Y. Zhang, Deep learning empowered task offloading for mobile edge computing in urban informatics. IEEE Int. Things J. **6**(5), 7635–7647 (2019)

81. Y. Mao, C. You, J. Zhang, K. Huang, K. Letaief, A survey on mobile edge computing: the communication perspective. IEEE Commun. Surv. Tutor. **19**(4), 2322–2358 (2017)

82. N. Cheng, W. Xu, W. Shi, Y. Zhou, N. Lu, H. Zhou, X. Shen, Air-ground integrated mobile edge computing networks: architecture, challenges, and opportunities. IEEE Commun. Mag. **56**(8), 26–32 (2018)

83. F. Zhou, R. Hu, Z. Li, Y. Wang, Mobile edge computing in unmanned aerial vehicle networks. IEEE Wirel. Commun. **27**(1), 140–146 (2020)

84. N.H. Motlagh, M. Bagaa, T. Taleb, UAV-based IoT platform: a crowd surveillance use case. IEEE Commun. Mag. **55**(2), 128–134 (2017)

85. M. Hua, Y. Wang, Q. Wu, C. Li, Y. Huang, L. Yang, Energy optimization for cellular-connected multi-UAV mobile edge computing systems with multi-access schemes. J. Commun. Netw. **3**(4), 33–44 (2018)

86. X. Cao, J. Xu, R. Zhang, Mobile edge computing for cellular-connected UAV: computation offloading and trajectory optimization, in *Proceedings of the IEEE International Workshop on Signal Processing Advances in Wireless Communications*, Kalamata, Greece, June, 2018

87. M. Messous, S. Senouci, H. Sedjelmaci, S. Cherkaoui, A game theory based efficient computation offloading in an UAV network. IEEE Transactions on Vehicular Technology **68**(5), 4964–4974 (2019)

88. S. Jeong, O. Simeone, J. Kang, Mobile edge computing via a UAV-mounted cloudlet: optimization of bit allocation and path planning. IEEE Trans. Veh. Technol. **67**(3), 2049–2063 (2018)

89. J. Zhang, L. Zhou, Q. Tang, E. Ngai, X. Hu, H. Zhao, J. Wei, Stochastic computation offloading and trajectory scheduling for UAV-assisted mobile edge computing. IEEE Int. Things J. **6**(2), 3688–3699 (2019)

90. F. Zhou, Y. Wu, R. Hu, Y. Qian, Computation rate maximization in UAV-enabled wireless powered mobile-edge computing systems. IEEE J. Sel. Areas Commun. **36**(9), 1927–1941 (2018)

91. X. Hu, K. Wong, Y. Zhang, Wireless-powered edge computing with cooperative UAV: task, time scheduling and trajectory design. IEEE Trans. Wirel. Commun. **19**(12), 8083–8098 (2020)

92. C. Zhan, H. Hu, X. Sui, Z. Liu, D. Niyato, Completion time and energy optimization in UAV-enabled mobile edge computing system. IEEE Int. Things J. **7**(8), 7808–7822 (2020)

93. J. Zhang, L. Zhou, F. Zhou, B. Seet, H. Zhang, Z. Cai, J. Wei, Computation-efficient offloading and trajectory scheduling for multi-UAV assisted mobile edge computing. IEEE Trans. Veh. Technol. **69**(2), 2114–2125 (2019)

94. J. Lyu, Y. Zeng, R. Zhang, Latency-aware IoT service provisioning in UAV-aided mobile-edge computing networks. IEEE Int. Things J. **7**(10), 10573–10580 (2020)

95. J. Lyu, Y. Zeng, R. Zhang, UAV-aided offloading for cellular hotspot. IEEE Trans. Wirel. Commun. **17**(6), 3988–4001 (2018)

96. X. Hu, K. Wong, K. Yang, Z. Zheng, UAV-assisted relaying and edge computing: scheduling and trajectory optimization. IEEE Trans. Wirel. Commun. **18**(10), 4738–4752 (2019)

97. M. Mozaffari, W. Saad, M. Bennis, M. Debbah, Efficient deployment of multiple unmanned aerial vehicles for optimal wireless coverage. IEEE Commun. Lett. **20**(8), 1647–1650 (2016)

98. J. Lyu, Y. Zeng, R. Zhang, T.J. Lim, Placement optimization of UAV-mounted mobile base stations. IEEE Commun. Lett. **21**(3), 604–607 (2017)

99. H. Chung, S. Maharjan, Y. Zhang, F. Eliassen, K. Strunz, Placement and routing optimization for automated inspection with UAVs: a study in offshore wind farm. IEEE Trans. Ind. Inf. **17**(5), 3032–3043 (2021)

100. H. Sun, F. Zhou, R. Hu, Joint offloading and computation energy efficiency maximization in a mobile edge computing system. IEEE Trans. Veh. Technol. **68**(3), 3052–3056 (2019)

101. X. Cao, J. Xu, R. Zhang, Mobile edge computing for cellular connected UAV: computation offloading and trajectory, in *Proceedings of the IEEE International Workshop on Signal Processing Advances in Wireless Communications* (IEEE, Kalamata, Greece, June 2018)

102. A. Filippone, *Flight Performance of Fixed and Rotary Wing Aircraft* (American Institute of Aeronautics and Astronautics, 2006)

103. National Centre for Earth Observation and National Centre for Atmospheric Science, The CEDA Archive: The Natural Environment Research Council's Data Repository for Atmospheric Science and Earth Observation. http://archive.ceda.ac.uk/

104. Kingfisher Information Services—Offshore Renewable Cable Awareness, Awareness chart of Walney 1–4. http://www.kis-orca.eu/downloads

105. AscTec Falcon 8. http://www.asctec.de/en/uav-uas-drones-rpas-roav/asctec-falcon-8/pane-0-1

106. F. Zhou, Y. Wu, R. Hu, Y. Qian, Computation rate maximization in UAV-enabled wireless powered mobile-edge computing systems. IEEE J. Sel. Areas Commun. **36**(9), 1927–1941 (2018)

107. S. Nakamoto, *Bitcoin: A Peer-to-Peer Electronic Cash System* (2008). https://bitcoin.org/bitcoin.pdf

108. S. King, *Primecoin: Cryptocurrency with Prime Number Proof-of-Work* (2013). http://primecoin.io/bin/primecoin-paper.pdf

109. A. Miller, A. Juels, E. Shi, B. Parno, J. Katz, *Permacoin: Repurposing Bitcoin Work for Data Preservation* (IEEE Symposium on Security and Privacy, San Jose, CA, May 2014), pp. 475–490

110. Slimcoin: A peer-to-peer crypto-currency with proof-of-burn. https://eprint.iacr.org/2019/1096.pdf

111. S. King, S. Nadal, *PPcoin: Peer-to-Peer Crypto-Currency with Proof-of-Stake* (2012). https://peercoin.net/assets/paper/peercoin-paper.pdf

112. A. Kiayias, A. Russell, B. David, R. Oliynykov, Ouroboros: A provably secure proof-of-stake blockchain protocol, in *37th Annual International Cryptology Conference*, Santa Barbara, CA, 20–24 Aug 2017 (IEEE, 2017), pp. 357–388

113. I. Bentov, C. Lee, A. Mizrahi, M. Rosenfeld, Proof of activity: extending Bitcoin's proof of work via proof of stake (extended abstract). ACM SIGMETRICS Perform. Eval. Rev. **42**(3), 34–37 (2014)

114. Z. Zheng, S. Xie, H.N. Dai, X. Chen, H. Wang, Blockchain challenges and opportunities: a survey. Int. J. Web Grid Serv. **14**(4), 1–25 (2018)

115. D. Larimer, *Delegated Proof-of-Stake (DOPS)* (Bitshare whitepaper, 2014)

116. EOS.IO, Technical White Paper v2 (2018). https://github.com/EOSIO/Documentation/blob/master/TechnicalWhitePaper.md

117. M. Castro, B. Liskov, Practical Byzantine fault tolerance, *USENIX Symposium on Operating Systems Design and Implementation(OSDI)*, vol. 99 (1999), pp. 173–186

118. Hyperledger project (2015). https://www.hyperledger.org/

119. X. Zhang, J. Liu, Y. Li, Q. Cui, X. Tao, R.P. Liu, Blockchain based secure package delivery via ridesharing, in *IEEE 11th International Conference on Wireless Communications and Signal Processing, Xi'an*, 23–25 Oct 2019 (IEEE, 2019), pp. 1–6. https://doi.org/10.1109/WCSP.2019.8927952

120. Y. Dai, D. Xu, K. Zhang, S. Maharjan, Y. Zhang, Deep reinforcement learning and permissioned blockchain for content caching in vehicular edge computing and networks. IEEE Trans. Veh. Technol. **69**(4), 4312–4324 (2020). https://doi.org/10.1109/TVT.2020.2973705

121. J. Kang, R. Yu, X. Huang, S. Maharjan, Y. Zhang, E. Hossain, Enabling localized peer-to-peer electricity trading among plug-in hybrid electric vehicles using consortium blockchains. IEEE Trans. Ind. Inf. **13**(6), 3154–3164 (2017)

122. H.B. McMahan, E. Moore, D. Ramage, S. Hampson, B.A.Y. Arcas, Communication-efficient learning of deep networks from decentralized data, in *Proceedings of the 20th International Conference on Artificial Intelligence and Statistics*, Fort Lauderdale, Florida, Apr 2017, vol. 54 (PMLR, 2017)

123. Y. Lu, X. Huang, K. Zhang, S. Maharjan, Y. Zhang, Low-latency federated learning and blockchain for edge association in digital twin empowered 6G networks, in *IEEE Transactions on Industrial Informatics* (In Press). https://doi.org/10.1109/TII.2020.3017668

124. A. Hard, K. Rao, R. Mathews, F. Beaufays, S. Augenstein, H. Eichner, C. Kiddon, D. Ramage, *Federated Learning for Mobile Keyboard Prediction* (2018). arXiv:1811.03604

125. Z. Yu, J. Hu, G. Min, H. Lu, Z. Zhao, H. Wang, N. Georgalas, Federated learning based proactive content caching in edge computing, in *2018 IEEE Global Communications Conference*, Abu Dhabi, United Arab Emirates (IEEE, Piscataway, 9–13 Dec 2018), pp. 1–6

126. J.H. Mills, G. Min, Communication-efficient federated learning for wireless edge intelligence in IoT. IEEE Int. Things J. **7**(7), 5986–5994 (2019)

127. H.B. McMahan, E. Moore, D. Ramage, S. Hampson, B.A.Y. Arcas, Communication-efficient learning of deep networks from decentralized data, in *Proceedings of the 20th International Conference on Artificial Intelligence and Statistics*, Fort Lauderdale, Florida, vol. 54 (PMLR, 2017), pp. 1273–1282

128. S. Wang, T. Tuor, T. Salonidis, K.K. Leung, C. Makaya, T. He, K. Chan, Adaptive federated learning in resource constrained edge computing systems. IEEE J. Sel. Areas Commun. **37**(6), 1205–1221 (2019)

129. C. Dinh, N.H. Tran, M.N.H. Nguyen, C.S. Hong, W. Bao, A.Y. Zomaya, V. Gramoli, Federated learning over wireless networks: convergence analysis and resource allocation. IEEE/ACM Trans. Net. **29**(1), 398–409 (2021). https://doi.org/10.1109/TNET.2020.3035770

130. S.J. Pan, Q. Yang, A survey on transfer learning. IEEE Trans. Knowl. Data Eng. **22**(10), 1345–1359 (2010). https://doi.org/10.1109/TKDE.2009.191

131. Y. Liu, Y. Kang, C. Xing, T. Chen, Q. Yang, A secure federated transfer learning framework. IEEE Intell. Syst. **35**(4), 70–82 (2020). https://doi.org/10.1109/MIS.2020.2988525

132. H. Yang, H. He, W. Zhang, X. Cao, FedSteg: a federated transfer learning framework for secure image steganalysis, in *IEEE Transactions on Network Science and Engineering* (In Press). https://doi.org/10.1109/TNSE.2020.2996612

Printed in the United States
by Baker & Taylor Publisher Services